MW01011857
WHOLE
FARM
MANAGEMENT

FROM START-UP TO SUSTAINABILITY

EDITED BY
GARRY STEPHENSON

CONTRIBUTING AUTHORS
Nick Andrews, David Chaney, Melissa Fery, Amy Garrett,
Lauren Gwin, Melissa Matthewson, Tanya Murray, Heidi Noordijk,
Sherri Noxel, Maud Powell, Garry Stephenson, and Josh Volk

Storey Publishing

The mission of Storey Publishing is to serve our customers by publishing practical information that encourages personal independence in harmony with the environment.

Edited by Deborah Burns, Sarah Guare, and David Chaney
Art direction by Carolyn Eckert
Book design by Carolyn Eckert and Erin Dawson
Indexed by Christine R. Lindemer, Boston Road Communications

COVER CREDITS
Cover photography by © **Lisa J. Godfrey**, front (t.r); © **Shawn Linehan**, front (t.l. & b.), back (l.), spine; © **Carmen Troesser**, back (r.)

INTERIOR PHOTOGRAPHY BY
© **Carmen Troesser**, 5, 8, 10 b.l., 13 b., 19 t.r., 29 r., 42, 62, 79 r., 83 t., 102, 103, 129–131, 220, 223, 227 l., 233 r., 237 r., 238 t.r., 248 r., 256 t., 261, 266 l., 303
© **Lisa J. Godfrey**, 10 t.r., 11, 14, 16, 18 b.l., 29 l., 31, 33, 34, 51, 60 r., 90, 92, 93, 113 t., 137, 157 r., 160 r., 176 l., 178, 200 r., 225, 228, 237 l., 256 b., 310
© **Lise Metzger**, 9 l., 10 t.l., 19 m.r., 29 c., 39 r., 55 t., 84 l., 98 t.l., 104 t., 108, 109, 132, 133, 139 l., 173 r., 232, 245 t.r., 263, 266 t.r.
© **Shawn Linehan**, 1, 2, 7, 9 all but l., 10 t.c. & b.r., 13 t., 15, 17, 18 all but b.l., 20, 22, 25, 38, 39 c., 43–46, 48, 50, 53, 54, 55 b., 56 t., 57, 59, 61 t., 63, 65 r., 67, 68, 70, 71, 73 l., 76–78, 79 l., 81, 83 b., 84 r., 85, 88, 89 all but b.r., 94–96, 98 all but t.l., 101, 104 all but t., 110, 112, 113 b., 114, 116, 117, 119 l. & c., 120, 121, 123, 126, 127, 134, 139 c. & r., 142, 146 l., 153, 154, 155 l., 157 l., 160 l., 161, 168–172, 173 l. & c., 174, 175, 176 c., 177, 184–187, 197–199, 200 l., 203, 212, 215, 219, 222, 224, 226, 227 r., 233 l. & c., 234–236, 238 all but t.r., 240–243, 245 t.l. & t.c., 246, 247, 248 l., 249, 251–255, 258, 262, 266 b.r., 305, 306, 309

ADDITIONAL PHOTOGRAPHY BY
© Brandi Larkin/stock.adobe.com, 122; © Eric Buell/stock.adobe.com, 190 r.; © erika_mondlova/stock.adobe.com, l.; © Deborah Vernon/Alamy Stock Photo, 193 r.; © fotokostic/iStock.com, 39 l.; © Garry Stephenson, 47; © itsajoop/iStock.com, 69; © Julia Duke Photo, 19 t.l.; © Jürgen Fälchle/stock.adobe.com, 87; © Kevin Miller/iStock.com, 72 r.; © laurha/stock.adobe.com, 193 c.l.; © Mark Ylen, Albany-Democrat-Herald, 19 b.l., 56 b., 216, 217; Courtesy of Maxwell Joel Cohen, 60 l.; © Modfos/iStock.com, 72 l.; © ne_fall_photos /stock.adobe.com, 106; © Nigel Cattlin/Alamy Stock Photo, 193 c.r.; © Noel Bridgeman/Alamy Stock Photo, 61 b.; © Oregon State University, 19 m.l., 73 r., 119 r., 176 r., 214; © randimal/stock.adobe.com, 190 c.; Courtesy of Sarahlee Laurence, 65 t.l.; © simona/stock.adobe.com, 190 l.; © Stephen Smith, 218, 245 b.; © Sweet Home Farms, 89 b.r., 143, 146 r., 155 r., 210, 211; courtesy of USDA NRCS, 41 l., 74, 75; © Vera Kuttelvaserova/stock.adobe.com, 189; © YinYang/iStock.com, 65 b.l.
Logos page 107 courtesy of (top to bottom): USDA, Food Alliance, Humane Farm Animal Care, Salmon-Safe, The Non-GMO Project, Fair Trade Certified
Farm map illustration page 41 Courtesy of Jen Keller, Pacific Crest Farm
Graphics by Ilona Sherratt, adapted from Oregon State University Professional and Continuing Education (PACE) Growing Farms: Successful Whole Farm Management: 22, 66, 101, 179, 194; USDA NRCA, 36 and sources credited on pages 181, 183, 185, 191, 192

Storey Publishing
210 MASS MoCA Way
North Adams, MA 01247
storey.com

Printed in Malaysia through Asia Pacific Offset
10 9 8 7 6 5 4 3 2 1

Library of Congress Cataloging-in-Publication Data

Names: Stephenson, Garry Owen, editor.
Title: Whole farm management : from start-up to sustainability / edited by Garry Stephenson ; contributing authors: Nick Andrews, David Chaney, Melissa Fery, Amy Garrett, Lauren Gwin, Melissa Matthewson, Tanya Murray, Heidi Noordijk, Sherri Noxel, Maud Powell, Garry Stephenson, and Josh Volk.
Description: North Adams [Massachusetts] : Storey Publishing, 2019. | Includes bibliographical references and index. | Summary: "Whole Farm Management is a comprehensive guide developed by the Small Farms Program at Oregon State University to help aspiring and beginner farmers make smart, strategic business decisions to ensure lasting success"— Provided by publisher.
Identifiers: LCCN 2019025786 (print) | LCCN 2019025787 (ebook) | ISBN 9781635860740 (paperback) | ISBN 9781635860764 (hardcover) | ISBN 9781635860757 (ebook)
Subjects: LCSH: New agricultural enterprises. | Farm management. | Farms, Small.
Classification: LCC S565.86 .W46 2019 (print) | LCC S565.86 (ebook) | DDC 630.68—dc23
LC record available at https://lccn.loc.gov/2019025786
LC ebook record available at https://lccn.loc.gov/2019025787

ACKNOWLEDGMENTS

It is important to acknowledge those who have financially and enthusiastically supported our work over the years, allowing us to develop and teach the curriculum that is published here. Oregon Tilth, Inc., through its partnership with the Center for Small Farms & Community Food Systems, has supported our work with beginning farmers, including the development of the course Growing Farms, since 2009. The USDA's National Institute of Food and Agriculture (NIFA) Beginning Farmer and Rancher Development Program has provided grant support to advance our work in beginning farmer education.

All royalties from this book will support the Oregon State University Center for Small Farms & Community Food Systems and its work in sustainable agriculture and community food systems.

CONTENTS

SELL
IT

MANAGE
IT

GROW
IT

KEEP
IT

INTRODUCTION

GROWING A FARM

MANY PEOPLE DREAM of starting a farm. If you ask them "Why a farm?" you will get a wide variety of responses. And if you ask "What kind of farm?" the answers will cover a range of possibilities. Some people describe their dream farm in great detail, and others describe it in broad strokes. This is hardly surprising: there are probably as many types of farms and farm businesses as there are unique personalities.

Yet across these diverse dreams and possibilities, there are basic principles that apply in every situation. Operating a farm business requires managing dreams, crops, people, markets, money, and reality.

How to manage all six of those — individually, then together — is the subject of this book. In the following chapters, we lead aspiring and beginning farmers through the basics of how to start and manage a successful farm business. In the process, we blend advice and inspiration from experienced farmers with guidance from agricultural educators. The farmers we work with, like those who are the focus of this book, operate small to medium-sized farms that use sustainable and organic methods, and they sell their products through local and regional markets.

What do we mean by "aspiring and beginning farmers"? This book is intended for people who are thinking of starting a farm business (particularly a small farm), those within their first decade of farming, and others who are not beginners but are considering major changes to their farm business. If you do not fit into any of these categories, this book is still a rich resource, with information you can adapt to any farming situation. If you are not planning to have your own farm and would rather be a farm manager for someone else, this book is also for you.

Where you are in this whole process will determine how you use this book. If you are not farming yet and are in the early stages of imagining your farm, it will be useful to go through the chapters in order. At the other end of the spectrum, if you are an experienced farmer, you might start with the table of contents, scanning for information relevant to your particular situation and the questions you have. One thing we have learned from farmers over the years is that their readiness for absorbing different types of information depends on what stage of farm business development they are in. (More on that later.)

THE BIG PICTURE

This book is based on a course developed specifically for beginning farmers. The course, Growing Farms: Successful Whole Farm Management, has been taught for the past 10 years in various locations throughout the state of Oregon and is now available online to reach a wider audience. In recent years, more and more people have been drawn to farming as a career that allows them to combine their

occupational and personal life goals. In the Growing Farms course, and in this book, we share valuable information and experiences that can help you get started farming, grow your farm business, and keep on farming into the future.

We have been at this long enough, however, to know starting and managing a farm is not easy, and many farms (like many small businesses) do not succeed. Farming is challenging, complicated, and sometimes deeply frustrating — both as a business and as a way of life. Some of the hardest things about farming are beyond a farmer's direct control, such as the high price and limited availability of appropriate farmland, tough marketplace competition, and thin profit margins. Having a clear-eyed perspective about these realities is an important part of managing your farm business. In fact, after reading this book, you might decide not to pursue farming — a positive step that may spare you hardships.

WHAT IS THIS BOOK ABOUT?

This book is not about how to grow specific crops or raise livestock (although we review the natural processes associated with farming). It is about how to grow a farm — the basic principles of starting and managing a farm business. As such, it is a great place to start for aspiring and beginning farmers or any farmer implementing changes on his or her farm.

Our discussion is grounded in three central ideas:

- **Managing:** *Whole farm management* provides the framework for developing a successful farm or ranch business.
- **Learning:** Farmers just starting out go through four predictable stages of development, and the stage individual farmers are in affects how and what and when they learn.
- **Succeeding:** Small farmers define success in many different ways and find fulfillment in *all* the opportunities a farm provides.

Mimo Davis and son, August, make the rounds at Urban Buds farm.

WHAT IS THE WHOLE FARM?

The whole farm is larger and more complex than you think. Clearly, the immediate focus is the physical farm (the land and everything that happens there), but it does not end at the farm gate. It ultimately involves relationships beyond the farm's boundaries. For example, with:

- Neighbors
- Direct market customers
- The community
- Wholesale buyers
- Processors
- Suppliers
- Competitors
- Land-grant university Cooperative Extension staff
- State and federal agencies
- Your banker

We emphasize whole farm planning as a road map for developing your whole farm management skills. Whole farm planning helps you identify the many connections and interactions within the farm and develop an integrated approach to decision-making that considers all the components of the farm business — environmental, economic, and social. Whole farm planning is farmer-defined and farmer-driven, and the process will be different for each farmer and farm. To help you on your way, we include a template for developing your own whole farm plan in the appendix. Keep in mind that whole farm planning is an ongoing process and something you will need to continually revisit as your farm business grows and develops.

Managing: What Is Whole Farm Management?

Whole farm management is a holistic and progressive approach to farm management. The basic premise is that you can make better decisions when you look at the whole farm. From this perspective, farmers take into account all the different components within the farm system and how they interact:

- **Physical resources:** Soil, water, and climate resources that are the physical base of the farm, plus the compatible infrastructure added in terms of farm equipment, irrigation, and so on
- **Biological resources:** Soil organisms, beneficial insects, wildlife, crops, livestock, pests, and more
- **Financial resources:** Business planning, markets and marketing, accounting systems, liability, profitability, and more
- **Human resources:** The household or human dimension of the farm, the package of skills and source of labor, and the desire for work that is satisfying and fulfilling

"You can't know it all at once. **Growing is where it starts. You have to know you can grow something before you can figure out how to sell it. But once you grow it, you have to figure out how to market. And once you've sold a few things, then you're in a position to ask: Can I afford to keep doing this?"**

— SMALL-SCALE GRAIN FARMER, OREGON

Managing the farm includes an awareness of all the connections between the farm's internal and external components. Decisions in one part of the farm affect other parts of the farm directly or indirectly. This dynamic plays out in pragmatic issues, such as matching the correct size main line pipe to an irrigation pump so *all* the peppers can be irrigated during a hot spell right after planting.

BEGINNING FARMER STAGES OF DEVELOPMENT

STAGE	YEARS	DESCRIPTION	WHAT FARMERS SAID AT EACH STAGE
1	1 to 3	**Proving we can grow and sell:** basic farm and crop management and marketing	"We didn't know enough to know how much we didn't know." "I skipped the business planning part and just started growing things."
2	3 and 4	**More deliberate, less frantic:** ease the burden, invest in and fine-tune infrastructure and marketing	"We are farmers now." "I have to be bent over for 2½ hours straight?"
3	4 to 6	**How to make money:** focus on business management	"We were excited to grow and sell . . . then we started to look at the numbers." "How do we scale up profitably without burnout?"
4	5 to 7+	**Big-picture thinking:** "We can do it. Should we still?"	"This is going to evolve . . . may need to try a couple of times." "Year 5 was a breakthrough, balancing the stresses of what we're trying to do."

It also plays out in market and production issues. For example, if you plan to increase production to meet market demand, it is crucial to consider the impact on other parts of the farm system — the *whole farm*. Will increasing production lead to overgrazing pastures? Will it shorten rotations and intensify vegetable pest management and soil health problems? Does the farm have the capacity in terms of processing, wash and pack, cooler space, labor, and other infrastructure to support increasing production? If the answers are no, then there could be profound negative effects on the farm's long-term productivity. If the answers are yes, then the whole farm is ready, and increasing production will benefit the whole farm.

Randy Kiyokawa gives a tour of Kiyokawa Family Orchards.

Learning: Farmers Have Developmental Stages

We have been working with beginning farmers and ranchers for more than 20 years. Training new farmers and helping them launch their businesses has long been a national priority, and both federal and state public agencies have dedicated programs and funding to make this happen. Yet, even with a successful start, we also see many beginning farmers decide to leave farming within their first 10 years.

Certainly, this is true for many career paths: even when people are successful in a career, many decide to change direction and do something new. From our research, we have learned that farmers and their farm businesses tend to move through four fairly predictable stages of development during those first 10 (or so) years. This matters because farmers are "ready" for different information at different times. Education and training programs need to meet new farmers where they are on their path.

How do we know this? After so many years of working with farmers, we decided to step back and study how farmers develop over time. We engaged with experienced and beginning farmers over the course of three years

Miranda Duschack works with the beehives at Urban Buzz, an enterprise of Urban Buds.

Anton Shannon and Rocco the dog lead the team to work at Good Work Farm.

(with surveys, focus groups, and interviews). When we analyzed all of the data, we found a clear pattern (see Beginning Farmer Stages of Development table, page 12). Farmers vary in how many years they spend at each stage, but they go through the stages in order.

As they pass through these stages, farmers confront how to produce crops, assess markets, plan and improve infrastructure, hire and manage labor, establish and use increasingly complex accounting systems, balance family and farm demands, and much more. Again, their need for information and training on these topics is to a large degree determined by their stage of development as farmers and business operators.

Why all of this matters. Beginning farmers and ranchers cannot learn everything immediately, do not always know what they need to know (as they themselves have reflected years later), and are motivated to find information and training when they encounter specific questions and challenges related to their particular stage of development. As you go through this book, you will probably absorb some parts and skim others. The beauty of having this book is that you can come back later, when you are ready for the parts you did not or could not use immediately.

We have observed that new farmers are often overly confident about what they think they know and what they can do, and that this attitude shifts as they gain more experience. This is simply worth keeping in mind. Initial overconfidence may even be an advantage in charging up the steep learning curve.

For example, a farmer in his tenth year of farming told us that when he first took our course Growing Farms, he felt confident about what he needed to know and ignored the rest. It was only after several years' farming that he realized what he had missed and wished he had listened to all of it.

Yet this is exactly our point: He simply wasn't ready to learn certain things until he had been farming for a few years. Only then did he have a framework to absorb and use that information. Now he is even more open and ready for new ideas, knowledge, and skills. "I am astonished every day at how much more I need to learn. People who come at farming from something else . . . learning what you have to learn . . . it does not end," he said.

Succeeding: How Do Small Farms Define Success?

We know that small farms provide an opportunity to design a business that is fulfilling and consistent with the goals of the farmers. Often, the conventional wisdom is that all businesses, including farm businesses, define success only in economic terms based on continuous growth. However, our research has revealed that although financial success is vital to small farmers, they are pursuing farming as more than just a way of making a living.

We learned from farmers that their visions of success reflect a variety of objectives they had incorporated into their farm businesses. Many develop their farms as a form of activism to facilitate social change or as an instrument to create a desired lifestyle or personal accomplishment. Farmers expressed the role of their farms as developing community; enhancing rural economies; supplying local, sustainable food; and preserving farm landscapes, genetics, and knowledge. Their comments clustered into four elements for defining success:

- **Social:** Farmers describe developing their businesses to create social change and be a positive part of their community.
- **Operational:** Farmers judge their success based on customer satisfaction and their reputation in the marketplace, and mastery of desired production techniques such as organic farming.
- **Lifestyle:** Farmers strive for work-life balance, physical and emotional health, and being at home with their family.
- **Financial:** Farmers set the predominant goal to be financially viable and cover household needs but not to maximize income at the expense of their values.

As you plan your farm, consider how these ways of defining success fit with your vision.

1: Dream It

In this chapter, you will consider the values that you bring to the business, build a vision and mission for your ideal farm, and create a foundation to move forward by setting some goals. As part of that exploration, you will begin to assess your farm business: the natural resources of your property that influence your choices about what to grow and produce, as well as the human resources — the personalities and skills of the people who will form your farm team. If you do not yet own or lease a farm, this chapter is a first step in organizing your thoughts around what you need or want in a farm before you buy — such as soils, water, and climate — and in beginning to create a plan for your future farm business.

2: Do It

This chapter is designed to help you think through the aspects of your farm that make it work for you and help you reach your goals. To use a computer analogy, it's about both the hardware (equipment and infrastructure) and software (people and processes) that make your farm run well. "Do It" will help you develop a holistic view of your farm operations and how to put resources to work: the practical aspects of equipment and infrastructure, irrigation, labor management, and farm energy.

3: Sell It

This chapter is based on the idea that before deciding what you want to grow on your farm, you need to know if and how you can sell it. You will be introduced to some basic marketing concepts and learn about the most common marketing channels used by small farms in the United States. Your mission statement, vision and values, and the other assessments you worked on in "Dream It" will play a major role in developing your marketing strategy.

4: Manage It

Sound business management is essential to the success of your small farm business. Business management does not need to be intimidating or overly complicated. With some

training and practice, all farmers can adopt successful business management practices. The purpose of this chapter is to help you get started in this learning process.

5: Grow It

"Grow It" is about the work of managing your land and resources as a farm ecosystem to produce crops and livestock. Given the diversity of farmers and ranchers and regional and climate variations, our goal is to introduce the physical and natural cycles that are linked to the basic principles and practices of sustainable agriculture. The chapter is not about how to grow specific crops; instead, it provides guidance that cuts across all types of farming systems — annual crops, perennial crops, and grass-based livestock systems.

6: Keep It

Now that you have a farm business, how do you keep it going for the long haul? There are challenges to operating any small business, and farms are no exception. Some might say that farm businesses are even more challenging given the number of variables that you have to deal with in producing, marketing, and selling your product, including your role as an entrepreneur, the major roles in a family business that will contribute to your satisfaction, the long-term success of the business, succession and estate planning, and more.

THE FARMERS

The 16 farmers featured in this book represent 9 farms from across the United States, producing a variety of crop and livestock products. Their experience covers a wide range of production systems, business models, and environments, and includes new start-up farms as well as multigeneration farms. As you will discover, each has a unique story in terms of how he or she has developed the farm business.

As you learn from them, keep in mind that their comments are from the moment in time when we interviewed them. These snapshots provide a rich picture of the challenges and rewards of starting a farm business. All the farmers included here continue to evolve and adapt with changing circumstances — developing new approaches to farming, changing the size of the farm, and finding new markets. One of them has transitioned out of farming, while another has developed a farming consulting business. Throughout this book, you will learn about how they got started and what drives them. Here is a quick introduction:

BLUE FOX FARM, Applegate Valley near Applegate, Oregon. **Operators:** Melanie Kuegler and Chris Jagger. **Products:** organic vegetables. **Size:** 67 acres. See profile on page 168.

GOODFOOT FARM, Kings Valley in the Coast Range near Philomath, Oregon. **Operators:** Beth Hoinacki and Adam Ryan. **Products:** organic and biodynamic vegetables and blueberries. **Size:** 10 acres. See profile on page 44.

GOOD WORK FARM, Lehigh Valley near Nazareth, Pennsylvania. **Operators:** Lisa Miskelly and Anton Shannon. **Products:** vegetables, fruits, and flowers from horsepower. **Size:** 14 acres. See profile on page 92.

KIYOKAWA FAMILY ORCHARDS, Hood River Valley near Parkdale, Oregon. **Operator:** Randy Kiyokawa. **Products:** apples, pears, peaches, cherries, plums, pluots, and pluerries. **Size:** 207 acres. See profile on page 94.

RAINSHADOW ORGANICS, in the high desert near Terrebonne, Oregon. **Operators:** Sarahlee Lawrence and Ashanti Samuels. **Products:** organic vegetables, grains, and livestock. **Size:** 130 acres. See profile on page 218.

SLOW HAND FARM, Sauvie Island, just outside of Portland, Oregon. **Operator:** Josh Volk. **Products:** organic vegetables. **Size:** ¼ acre. See profile on page 46.

SWEET HOME FARMS, foothills of the Cascade Mountains near Sweet Home, Oregon. **Operators:** Carla Green and Mike Polen. **Products:** grass-based beef, lamb, goat, and pork, and pastured poultry. **Size:** 82 acres. See profile on page 216.

URBAN BUDS: CITY GROWN FLOWERS, in the heart of downtown St. Louis, Missouri. **Operators:** Karen "Mimo" Davis and Miranda Duschack. **Products:** 70 varieties of cut flowers and honey. **Size:** 1 acre. See profile on page 129.

VANGUARD RANCH, near Gordonsville, Virginia. **Operators:** Renard and Chinette Turner. **Products:** goat meat and goat meat value-added items. **Size:** 94 acres. See profile on page 132.

Lastly, the 12 authors (listed on page 264) have drawn from their broad farming and professional experience to write this book. Together with the farmers listed here, we guide you to dream, do, sell, manage, grow, and keep your farm.

DREAM
IT
SLOW HAND FARM

CHAPTER ONE

STRATEGIC PLANNING

Starting and managing a farm business takes imagination. As you are hand-weeding that long row of garlic, tossing out hay for the livestock, or reflecting with your morning cup of coffee, it is important to dream about your farm business. In this chapter we encourage you to think about the values that you bring to the business, establish a vision and mission for your ideal farm, and set goals that will guide and inspire you as you move forward. To inform this creative process, you begin by working through an assessment of your farm business: the *human resources* — the personalities and skills of the people who form your farm team — as well as the *natural resources* of your property that influence your choices about what to grow. If you do not yet own or lease a farm, this is a first step in organizing your thoughts to begin creating a plan for your future farm business.

BY THE END OF THIS CHAPTER YOU WILL BE ABLE TO:

- Outline a whole farm planning process that helps you achieve your personal and business goals.
- Identify the vision and values for your farm business.
- Write a draft mission statement for your farm business.
- Develop an inventory and assessment of the natural and physical resources that describe your farm (or potential farm).
- Identify your farm team and work with that group of people to build your farm business.
- Create a draft SWOT analysis for your farm business.

WHOLE FARM PLANNING

As a farmer, you need to make many interconnected decisions. What you do for one aspect of the farm affects or has implications for other parts of the farm. For example:

- The goals and desires you share with your family and other business partners directly affect the kind of farm business and mission statement you create.
- The infrastructure and equipment you have on your farm influence your choices of what you can grow and your ability to keep crops productive through cold or hot weather.

When taking a whole farm perspective, farmers and ranchers have to manage, or at least be aware of, many different components within the farm system. The Whole Farm Components illustration (at right), with its various intersections and overlapping circles, shows how the different elements interrelate and affect each other.

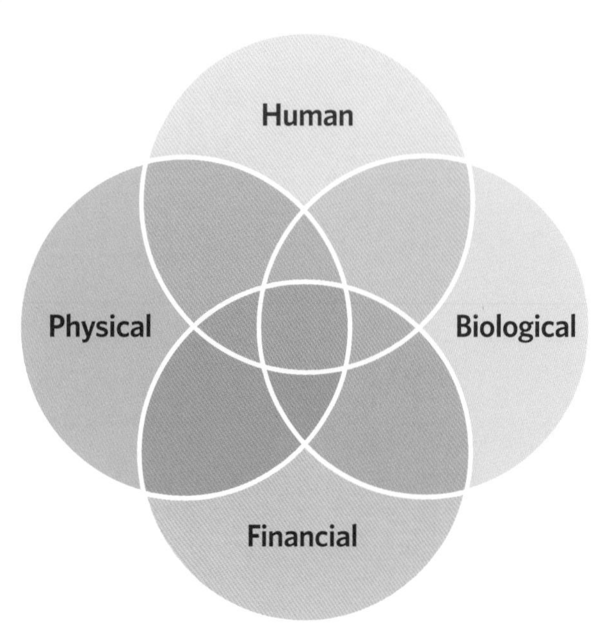

WHOLE FARM COMPONENTS

HUMAN

- Farm team and employees
- Seasonal labor
- Family members
- Customers
- Suppliers
- Advisors
- Local community

PHYSICAL

- Soil
- Water
- Land features
- Infrastructure
- Equipment

BIOLOGICAL

- Crops
- Livestock
- Soil organisms
- Pests
- Wildlife
- Natural vegetation

FINANCIAL

- Marketing
- Profitability
- Financial accounting
- Liability
- Taxes
- Payroll

In this book, we focus on whole farm planning as a strategy to deal with this complexity. Whole farm planning helps you identify the many connections and interactions within the farm and develop an integrated approach to decision-making that considers all the components of the farm business — environmental, economic, and social. A whole farm plan:

- Provides a road map for the future
- Helps you plan for future profitability
- Provides a foundation from which to expand and/or change your farm business
- Serves as a reference document
- Is a flexible, "living" process that should be regularly updated

To assist you in developing your own whole farm plan, we include a whole farm plan template — 23 separate worksheets (appendices 1-1 through 1-23) referenced throughout the book and identified by the Farm Plan icon (below). Where this icon appears, you will find a brief description of the relevant worksheet, plus instructions or suggestions for completing the form.

Farm plan worksheets are also available as Adobe PDF forms at **www.storey.com/farm-plan-worksheets.** You can download the PDF forms to your computer and then type directly into them and save them for your personal use. Full instructions are provided on the website. If you complete all the worksheets, you will have a basic whole farm plan that you can continue developing as you grow your business.

HUMAN RESOURCES

Let us begin by looking at the social and human side of owning and operating a farm business.

- Who is involved in the farm business?
- What are your skills, both individually and collectively?
- What lifestyle do you envision?
- What is your relationship to your surrounding community?

Tasks and Responsibilities in a Farm Business

To begin answering those questions, it is important to first consider what is actually involved in running a farm business. What tasks and responsibilities are on a farmer's to-do list? An initial list might include the following:

- Grow crops
- Raise animals
- Develop relationships with the surrounding community
- Purchase and maintain tools and equipment
- Market farm products and services
- Develop and update plans
- Access educational opportunities
- Comply with regulations
- Practice responsible land stewardship
- Maintain balance sheets
- Keep production and accounting records
- Hire and manage employees
- Finance farm operations
- Purchase seed, livestock, and inputs

For a more complete picture of what farmers do, review the assessment worksheets in *Exploring the Small Farm Dream*, published by the New England Small Farm Institute (see Resources).

If you are farming solo, making decisions about these tasks is ultimately up to you. However, if your farm business includes other family members or employees, their participation is going to be important. Including them in decisions — taking into account their ideas, thoughts, and feelings — can produce a better outcome overall and increase their buy-in and commitment to getting the work accomplished. This same principle applies as you develop your farm plan. Group decision-making is challenging, but the working relationships that develop over the long term will be worth it.

Each person on your farm team brings a unique set of skills, talents, and ideas to the business. The skills and preferences represented on your team will determine the answers to many business-related questions. For example:

- What will we produce on the farm?
- Can we purchase used equipment?
- What will be the best way to market our products?
- Will I need to pay for tax preparation?
- How much risk is our business willing to take on?

Your Lifestyle

In addition to assessing skills, talents, and preferences, it is important to think about the lifestyle you envision for yourself and your family. Owning and operating a farm offers a unique and rewarding life. Yet farming requires notoriously long hours and hard physical work, and farming can be a risky financial endeavor. In the whole farm planning process, be realistic about your farm business and verify that it is compatible with your life goals and dreams.

Exploring your quality of life is a subjective process aimed at evaluating how you want to be spending your time.

FARM TEAM SKILLS ASSESSMENT

Assess the skills and talents of your own farm team using the Farm Team Skills Assessment worksheet (appendix 1-1). Fill out the table for your entire farm team. You can meet as a group to work on this. Or you might print a blank copy for each team member and have each work on it and then meet to discuss your responses. Consider the following questions as you work on this: What are your skills and talents? What do you enjoy doing? What kind of work gets you excited? Think broadly, not just about farming per se. For example, if you are organized and enjoy working with numbers, you may be interested in taking on record-keeping responsibilities.

Column 1: List task categories and sample skills.

Column 2: Write in all the skills represented within your farm team.

Column 3: There will most likely be gaps in skills that you think are needed. Write down those skills that are missing in your farm team and add a brief phrase about how you plan to get that skill (e.g., get training, hire someone, contract work on as-needed basis).

If you want to work through another, more-detailed assessment, take a look at the *Northeast Small-Scale, "Sustainable" Farmer Skill Self-Assessment Tool* from the New England Small Farm Institute (see Resources).

QUALITY-OF-LIFE ASSESSMENT

The Quality-of-Life Assessment worksheet (appendix 1-2) encourages you to explore your own views about the quality of life you desire. Work through the series of questions presented in the worksheet. Have other members of your farm team answer these questions for themselves, then discuss your answers with each other to see if you can come to a consensus where there are differences of opinion. It is okay to leave sections blank if you cannot complete the form at this time. You can return to the worksheet later to make any changes or additions.

Key questions include:

- What do you want from your life?
- What would make farming enjoyable and rewarding for you, your whole family, or farm team?
- What hopes do you have for your family and for your community?
- What contributes to your health, happiness, and cultural/spiritual needs?

Social Capital

As we wrap up this discussion of human resources, think about the community in which you live and work. Farms, like rural schools, connect people in communities together. For many people, farms represent a set of values and a way of life. Farms can also be a focal point for social interaction and create a space where people come together around the common tasks of feeding ourselves and protecting the natural environment. These themes are all related to the concept of "social capital." Social capital refers to the collective value of all social networks within a region, and the tendencies that arise from these networks to do things that benefit each other. When you give to, and draw upon, the greater community, together you can solve common problems and strengthen the culture in your local area.

How do you begin building social capital for your own farm? First, consider what your farm can contribute to your community or region. Could your farm be used as a gathering place for a meeting or activity? Do you have surplus products at certain times of the year that could be donated? Are you an expert in a certain field in which you could share your knowledge?

Then work on the steps outlined below:

1. Assess resources you have to share with the community.
    - Farm property
    - Infrastructure
    - Products generated on the farm
    - You, the farmer
2. Get to know your neighbors and community.
3. Consider ways to work with other farmers.
    - Ask for help or advice.
    - Barter or trade.
    - Share equipment.
4. Create a social network among your customers.
5. Hold a social and/or educational event on your farm.

> To connect with customers, **you must go beyond just having a quality product to sell. You also need to create an identity for the business that draws people in and communicates who you are and what you have to offer them.**

CREATING AN IDENTITY FOR YOUR FARM

One of the most fundamental aspects of operating a farm business is building and maintaining your customer base. To connect with customers, you must go beyond just having a quality product to sell. You also need to create an identity for the business that draws people in and communicates who you are and what you have to offer them. This section highlights some of the key steps in this process.

Naming Your Farm

What is the name of your farm? Or if you are just in the initial planning phases, have you thought about what you would like to call your farm? The name you choose can communicate a lot about your farm business. A farm name can:

- Provide marketing opportunities.
- Describe what you do.
- Convey your mission/vision.
- Tell where you are located.
- Reflect your personality and values.

If you do not yet have a farm name, finding one that works well can be a challenge. The U.S. Small Business Administration offers the following suggestions:

- How will the name look? Will the name work well on a website, business cards, and advertisements or with a logo?
- How will it sound? Is it easy to pronounce?
- How will it be remembered? Are there positive connotations that help the name stick with customers? What makes your business unique? Choose a name that distinguishes you from competitors.

FARM NAME AND CONTACT INFORMATION

Fill out the Identifying Your Farm worksheet: Name and Contact Information (appendix 1-3). If you do not yet have a name for your farm, take some time to write down your ideas, then discuss them with other members of your farm team, colleagues, friends, or customers.

There are things to be careful of when choosing a farm name. First, avoid using embarrassing spellings, abbreviations, profanities, or potentially offensive undertones. Be careful about implied associations with organizations/people the business is not connected with. Finally, make sure you avoid trademark infringements.

As you work on your farm name, think about designing a logo at the same time. A logo is an important part of branding your business and enhances name recognition when it comes to marketing your product. Working on the name and logo in tandem can facilitate the decision-making and design process. Once you have some names in mind, check that the specific name is available and not already registered to an existing business in your state. To find the government department that deals with business registration and licensing, search online or check with your local library.

Defining Vision and Values

Your values are the core beliefs and philosophies that reflect your view on life. Whether you are aware of it or not, you bring those values to the tasks of starting and operating a farm business. Your values have a major impact on the goals you develop for your farm; they also help guide day-to-day farm management and will influence business decisions that you face in the future. Values typically do not change with time and are reflected in everything you do.

The list on page 29 presents 100 different words and terms that could be described as values. Some of these words might connect with you, some may have no meaning or have a neutral connotation, while others may elicit a negative response. The words represent professionalism, passion, ethics, hope, relationships, nature, optimism, and more. Read through the

SAMPLE VISION STATEMENTS

Our vision is to be a family farm business that is economically viable and socially responsible that produces unrivaled food using diversified, sustainable, and organic farming methods.
(Organic farm, California)

Our vision is to reflect the values and sustainability of a family farm while creating delicious, high-quality dairy products for everyone to enjoy.
(Small dairy, California)

We envision our farm within a sustainable food system, one in which farms are economically viable, supports our region's food needs, and feeds all members of our community. We believe our farm contributes to the health of our entire community through its social, economic, and environmental practices.
(Organic farm, Oregon)

list and write down the words that are meaningful to you — that reflect your values. After you have created your list, print it out and add any other values to the list that you feel are important to you.

Share your list with all the members of your farm team. Compare your lists with each other. Do you see similarities? How can your values be integrated to strengthen your farm business? Keep your printed list handy for filling out the Values and Vision worksheet (appendix 1–5).

Vision statement. What is your vision for your farm? What do you want your farm to look like 5, 10, 20 years down the road? A vision statement is a way of articulating those hopes and dreams in a few sentences. A vision statement is for you and your farm team, not for your customers or neighbors. It provides inspiration, establishes the big picture for all your business and strategic planning, and reminds you of what you are trying to build. There are so many decisions to make as you begin farming. Your vision statement helps you stay focused and is a reminder of why you are working so hard amid struggles and challenges.

A vision statement can address any aspect of the farm business that is important to you, such as:

- Type of farm, farming practices
- Which products you produce and market
- Community impacts
- Quality of life/lifestyle
- Environmental values
- Product quality values
- Employee and farmer welfare
- Profitability, fairness, and the like
- Providing customers with good food
- Level of income
- Contributions to charity
- What your inspiration is
- How you measure success

100 VALUE WORDS

- Adaptability
- Adventure
- Ambition
- Balance
- Beauty
- Belonging
- Camaraderie
- Caring
- Certainty
- Customer focus
- Education
- Effectiveness
- Efficiency
- Empathy
- Engaging
- Enthusiasm
- Excellence
- Experience
- Expertise
- Exploration
- Fairness
- Faith
- Family
- Fearlessness
- Financial independence
- Flexibility
- Fortitude
- Freedom
- Frugality
- Fun
- Generosity
- Gratitude
- Growth
- Happiness
- Harmony
- Health
- Honesty
- Hopefulness
- Humility
- Humor
- Imagination
- Independence
- Ingenuity
- Inquisitiveness
- Inspiration
- Integrity
- Joy
- Justice
- Kindness
- Leadership
- Logic
- Love
- Loyalty
- Mindfulness
- Modesty
- Nature/environment
- Neatness
- Open-mindedness
- Openness
- Optimism
- Order
- Originality
- Passion
- Peace
- Perceptiveness
- Perseverance
- Persistence
- Practicality
- Precision
- Professionalism
- Prosperity
- Punctuality
- Realism
- Reasonableness
- Reliability
- Resolve
- Resourcefulness
- Respect
- Rest
- Security
- Self-reliance
- Sensitivity
- Service
- Simplicity
- Sincerity
- Solitude
- Spirituality
- Spontaneity
- Stability
- Success
- Sustainability
- Teamwork
- Tradition
- Trust
- Truth
- Uniqueness
- Unity
- Usefulness
- Vision
- Wisdom

IDEAL FARM VALUES AND VISION

One way to help develop the vision for your farm is to describe or portray your ideal farm in some way. When you think about your current farm or dream about the one you plan to operate in the future, what do you picture? If you could have, do, grow, or raise anything and everything you want, what would be included? Fill out the Ideal Farm worksheet (appendix 1-4) to explore these questions. The form has a place for you to include a written description as well as space to add other ideas or illustrate this farm either through drawings or photos.

The Values and Vision worksheet (appendix 1-5) provides space for you to formulate a vision statement for the farm. This worksheet also has a place for you to summarize your values based on the list you created a few pages back.

When crafting your farm vision, let your imagination go and dream; let your vision statement capture your passion. The Sample Vision Statements box on page 28 provides several good examples of statements from farms.

What to Produce

The question about what to grow or produce on your farm is often the first thing that comes to mind when someone begins thinking about starting a farm business. "I'd really like to raise grass-fed beef." Or "I bet there's a lot of money to be made growing hops." If you are already farming, then you have worked through some of those decisions — although it is good to be open to other options. Others who are not as far along in the process, who are looking for land or who are trying to decide if farming is a good fit for them, may have only a general idea (or may not have any idea) of what they would like to produce.

Wherever you are along that continuum, approach your farm and business plan as an ongoing process. We encourage you to revisit your farm plan worksheets after you have evaluated your natural resources and as you work through other chapters in this publication. Part of the planning process is also recognizing that as you learn new information, the concept of your ideal farm may evolve and change. That's okay — planning and modifying your vision go hand in hand.

Developing a Mission Statement

Unlike the vision statement (a statement focusing on the future), the farm mission statement focuses on the current farm business. Your farm mission statement describes the overall purpose of your business and may include what you do, how and why do you do it, and whom you want to serve. Based on your vision and values, the mission statement

provides a concise summary of your farm business's purpose. You can use your mission statement as a marketing tool: share it on your website or advertising materials and when conversing with customers directly.

The mission statement usually includes specific, key information about your farm. Examples include:

- Customers who are served
- Products produced
- Skills and capabilities of your business
- Unique accomplishments

If possible, develop your mission statement as a farm team, with everyone involved. And periodically reexamine and update the mission statement to keep your business dynamic. You'll find ideas in the Sample Mission Statements box on page 32.

FARM PLAN

ENTERPRISE SELECTION

The three Enterprise Selection worksheets (appendices 1-6 through 1-8) will help you explore what you want to grow or produce on your farm. There are worksheets for Annual Crops, Livestock, and Perennial Crops. You don't need to complete them all if you already know the type of enterprise you want to establish on your farm; complete the worksheets that are most relevant to your situation. If you are completely undecided and open to looking into the whole range of animal and crop enterprises, then work through the questions for all three worksheets. This should bring some focus to the type of enterprise you want to develop.

SAMPLE MISSION STATEMENTS

We dedicate our business to the sustainable production of delicious specialty dairy products. We aspire to set the standard of excellence for quality in the products we produce, and we are committed to fairness and integrity in our partnerships with employees, customers, suppliers, and our other business relationships. We are devoted to being good stewards of the earth and responsible members of our community.
(Small dairy, northern California)

Our farm's mission is to produce great-tasting, high-quality beef using only grass, sunshine, and clean water the way nature intended.
(Grass-based ranch, Ohio)

Our mission is to provide certified organic produce and eggs to communities within our valley. We are committed to ecological farming principles and whole farm sustainability. We strive to educate consumers about the health benefits of organic vegetables and pastured poultry. We hope to be a model of profitability and sustainability.
(Organic farm, Oregon)

Setting Goals

If the mission statement communicates the overall purpose of your farm business, then goals describe *how* you are going to fulfill that purpose. Goal setting is a powerful process for deciding the direction in which you wish to take your business and for motivating the farm team to work toward that destination. To simplify the process, it may help to categorize goals as short, medium, and long term. If framed within a timeline, goals are more readily achievable. Once you have your list of goals, you can then prioritize them for better decision-making that can take your farm to the next level. Use the SMART acronym to help you develop meaningful and achievable goals:

- **Specific:** What do you want to accomplish and how?
- **Measurable:** How will you know when you have reached your goal? What indicators will you use?
- **Attainable:** Is the goal realistic and achievable? Do you have the skills, knowledge, and resources to reach it? Can you meet the goal without sacrificing too much?
- **Relevant:** What is the reason, purpose, or benefit of accomplishing the goal?
- **Timely:** When do you expect to reach your goal? Is that a reasonable time frame?

Following are two sample goal statements. Look for the SMART components in each goal.

Goal statement 1: By next November, increase farm profitability by 20 percent through a combination of increased yield, decreased expenses, and more competitive market pricing so that we can create a more successful business and reduce our level of stress.

Goal statement 2: Within five years, have a full-diet CSA that offers our members a wide diversity of products (including milk, meat, and eggs) that supply at least 75 percent of their caloric and nutrient needs in order to promote a healthy and enjoyable diet, enhance our farm's diversity and ecological health, establish stronger connections with our members, and increase the farm's profitability.

If writing out SMART goals seems too complicated, it is also acceptable to list simple statements of what you want to accomplish. For example:

- Make enough income on the farm to quit my off-farm job.
- Keep our children busy and earning money for a college fund.
- Actively build soil quality by cover cropping this fall.
- Grow enough product for two farmers' market booths next season.
- Finish 15 steers for direct sales this year.
- Build additional hay storage by next June.

As you begin a farm business, money and finances may be at the top of the list of your concerns. Farm finances are covered in more detail in chapter 4, "Manage It." In the meantime, here are key questions to think about relative to your financial goals:

- Does your farm need to provide the full income for your family or will it be supplemental to another job?
- How long can you wait before your agricultural enterprise generates income?
- What kind of capital can you invest up front?

FARM PLAN

MISSION AND GOALS

Refer to the Mission and Goals worksheet (appendix 1-9).

Mission statement: Before you start writing, take some time to brainstorm key ideas, themes, or phrases that you would like to include in your mission statement. Ideally, work on this brainstorming session with the other members of your farm team. Write all your ideas down on a separate piece of paper, then identify what's most important. Once you have agreed on those key themes and phrases, you can draft a few different mission statements in the space provided on the worksheet. Then discuss them with your farm team to come up with one that works best for your situation.

Goals: Write or type in your short-, medium- and long-range goals in the spaces provided.

NATURAL RESOURCES

In this section you will evaluate the physical and natural resources on your farm. Farms are not all created equal. Differences in soil and water resources have a huge impact on what can be grown or raised on a particular site. On some properties, you may be able to grow just about anything suited to your climate; on other land, you may be restricted to just a few options. This section provides information to help you answer the question "What can I grow on my farm?" It will also help you develop ideas for enterprises appropriate to your situation.

Soils

Soils are a living and dynamic component of all farm and ranch ecosystems. Soils support plant life; provide nutrients; and capture, store, filter, and release water. Managing the soil resource is one of your most important tasks. To do that well, you need to first understand the concept of *soil series classification*. Soil series classification groups soils based on similar chemical, physical, and formative properties. There are more than 18,000 soil series in the United States, and each type varies in depth, texture, and color. Most likely you have multiple soil series on your farm. Soil maps can give you an idea of the types of soils located in your general area. Walking your land, observing slight changes in topography and the way that water moves through the landscape, and feeling your soil are additional ways to identify differences between soils.

The USDA Web Soil Survey site (see Resources) can help you develop a more detailed understanding of the different soils on your farm. If you do not yet have land, or if you are considering purchasing or leasing a property, use the Web Soil Survey to explore those properties that you are interested in. In addition to identifying soil type, the Web Soil Survey provides information about the depth,

Lisa Miskelly and her horse team, Daisy and Duke, work a field at Good Work Farm.

structure, and restrictive features or limitations of soils. We recommend that you visit the Web Soil Survey site and create a soils map for your property. If you need assistance using the survey, contact your local Natural Resources Conservation Service office, soil and water conservation district, or Cooperative Extension Service.

Soil Texture

Soil texture is the composition of sand, silt, and clay particles in the soil. Texture is an inherent feature of the soil: for the most part, what you see is what you get. One acre-foot of dry soil weighs about 3.5 million pounds, so it is not practical to modify the soil texture at a field scale (for example, by adding sand). When you become familiar with the texture of your soil, you can manage the field more appropriately.

Soil texture is directly related to the soil characteristics that affect how water moves through the soil:

- Infiltration rate (water intake rate)
- Available water-holding capacity (amount of water in the soil that is available for plants)
- Permeability (ability of water to move through soil)

Texture plays a role in other important soil characteristics, such as ease of tillage, risk of erosion and compaction, soil fertility and quality, and seasonal accessibility.

The soil texture triangle (page 36) identifies the percentage of sand, silt, and clay in different soil textural classifications. If you know your soil textural classification, you can determine general percentages of sand, silt, and clay present in the soil. Alternatively, if you have the actual percentage of clay, sand, and silt, you can identify your soil textural classification by pinpointing where the three lines intersect.

LAND RESOURCE INVENTORY

A good place to start is to conduct a land resource inventory. Using the Land Resource Inventory worksheet (appendix 1-10), assess the general types of land on your farm. Identify the township, range, and section. Then for each type of land (tillable, pasture, woodlot, and so on), specify the number of acres, whether you own or rent that land, whether or not it has water rights, and any additional notes/questions you may have on production potential or other land use issues.

In the triangle, look at the differences between a soil that has a high clay content (such as a silty clay) and a soil with fairly equal components of sand, silt, and clay (such as a loam). In the loam soil, water infiltrates and drains more quickly than in the clay soil, resulting in improved soil aeration and a more optimal environment for root growth and development. The silty clay soil requires less irrigation since it retains more water. When given these two choices, farmers growing irrigated row crops would prefer the loam soil, while ranchers raising livestock on unirrigated pasture would prefer the silty clay.

The percentage of sand, silt, and clay can be determined in a soil lab, or it can be estimated by feel. The video *Determining Soil Texture by Hand* and the flow chart *Estimating Soil Texture*, from Washington State University, cover simple methods that farmers can use in the field to help determine soil texture (see Resources).

USING THE SOIL TEXTURE TRIANGLE

First, if you already know the percentages of clay, sand, and silt, locate the percent clay for your soil on the left edge of the triangle and the percent sand on the bottom edge. Follow the grid lines for those two numbers. The point where the two lines meet indicates the textural classification. For example, if your soil is 20 percent clay and 40 percent sand, then you have a loam soil and you can assume that the remaining 40 percent is silt. You can check that the percent silt agrees with your lab test results. USDA NRCS offers a Soil Texture Calculator to help you with this assessment (see Resources).

Second, look at the soil map you created using the Web Soil Survey and identify the soil series on your farm. Identify the texture of the soil from the name. For example, in the soil types called Jory silty clay loam or Newburg fine sandy loam, the terms "silty clay loam" and "fine sandy loam" refer to soil texture.

Once you have identified your soils from the Web Soil Survey, look for that description in the soil texture triangle and record the ranges for percent sand, silt, and clay.

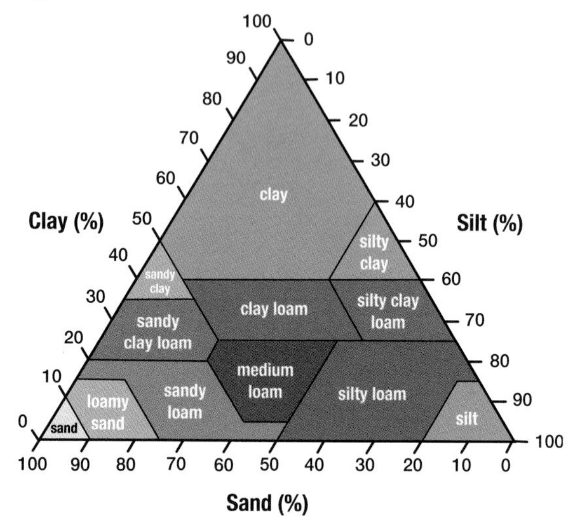

SOIL TEXTURE TRIANGLE

Soil Capability Classes

In addition to understanding your soil type, another aspect of evaluating land is to identify its capability class. There are eight capability classes that help define the land's production potential, historical usage, and limitations. Several factors are taken into consideration when classifying land this way, including slope, drainage, and available water-holding capacity of the soil (all linked to soil type).

Class I: Prime farmland

Class II: Good farmland with moderate limitations

Class III: Soils have severe limitations that reduce the choice of plants or require special conservation practices, or both

Class IV: Soils have very severe limitations that restrict the choice of plants or require very careful management, or both

Classes V and VI: Limitations are impractical to remove, unsuitable for cultivation, possible pasture, woodland, or wildlife areas

Classes VII and VIII: Mainly forestland and wildlife areas

One additional step is to determine the soil capability subclass. Soil groups within one subclass are identified by their limiting characteristic:

- **e (erosion):** Soils for which erosion is the main challenge or hazard affecting their use
- **w (water):** Soils for which excess water, through poor drainage, wetness, a high water table, or overflow, is the main challenge or hazard affecting their use
- **s (shallow, droughty, or stony):** Soils that are limited for use by the rooting zone, through shallowness, stone, or low intrinsic water- or nutrient-holding capacities
- **c (climate, too cold or too dry):** Soils for which temperature or moisture are the main limiting factors affecting their use

To get a better sense of why these soil characteristics are important, consider the following two soil types: silty clay loam (Class IIIw) and sandy loam (Class I). Which soil do you think would be best for growing crops? Which soil do you think would be best for unirrigated pasture? Why?

Sandy loam (Class I) would be the best choice for growing vegetable crops. A Class I sandy loam has good tilth and good water-holding capacity and also drains well, so it provides an optimal environment for root development.

Silty clay loam (ClassIIIw) would be the best for unirrigated pasture. With a higher water-holding capacity (due to higher clay content), this soil can provide the pasture with the water it needs over a longer period. Also, pasture plants can tolerate the excess water and poor drainage that exists on this soil (vegetable crops would not tolerate these conditions).

In summary, soil is a valuable and important resource. It is literally the foundation of your farm. Soils are highly variable, so it is critical that:

SOILS ASSESSMENT

Building on the big-picture assessment you developed with the Land Resource Inventory worksheet, refer to the Soils Assessment worksheet (appendix 1-11) to assess the soil types on your farm. For each soil, include details about capability class, drainage, erosion risk, and other key characteristics.

1. You know your particular soil.
2. You are able to define its characteristics.
3. You understand how best to manage it.

In this chapter, we have looked at the first two items in that list. You will learn more about soil management in chapter 5, "Grow It."

Water

Having access to adequate irrigation water gives you more options for crops that can be grown on your farm. How farmers obtain water for irrigation varies by region and climate. It may include predictable rainfall during the growing season, groundwater from wells, and surface water from lakes, reservoirs, ponds, springs, streams, and rivers. It is also important to look carefully at both water quality and — when too much or too little are a problem — water quantity.

Water access. Laws regarding access to water also vary from state to state, each with its own unique regulations pertaining to water allocation. In regions with lower annual precipitation — such as most western U.S. states — water use is closely regulated. We use

FARM PLAN

WATER RESOURCES ASSESSMENT

To help you evaluate water resources on your own farm, refer to the Water Resources Assessment worksheet (appendix 1-12). For each source of water on your farm, describe how it will be used, the water delivery system you will use to get water to the land, and any limitations you are aware of related to water quantity, quality, or other issues.

Oregon here as a case study to illustrate the types of issues you may need to consider in obtaining water for your farm. For information about water rights and regulations in your own state, contact your state's water resources department.

Balancing water needs for irrigation, livestock, wildlife, and recreation is a challenge in many farming communities. Under Oregon law, water is publicly owned and, with a few exceptions, farmers and other users must obtain a "water right" or a permit from the state to access that water. The law is based on the concept that water is a public resource with multiple uses; therefore a private individual cannot "own" it. In Oregon, the exempted uses from water rights are:

- Domestic wells
- Livestock watering
- Fire control
- Springs (that develop and reenter the soil on the same land)
- Rainwater collected from roofs or other impermeable surfaces

Most surface water in Oregon is appropriated for specific uses, and a large portion of that is designated for agriculture. In some areas, irrigation districts are also involved in the allocation and use of water. Irrigation districts are self-governing, with elected board members who manage the irrigation project and set policy regarding distribution of water in the district. Groundwater that is used for irrigation will be from a designated irrigation well. Such wells tend to be deep, high-volume wells that require permits (prior to drilling) and must be registered with the state.

During drought periods when surface water is limited, the concept "first in time, first in right" is used to determine who has priority access to surface water. Under Oregon law, land with older, more senior water rights has access to the water first. The properties with the oldest water rights can take the quantity they are legally allotted first, before properties that have had water rights granted more recently. If farmers are looking to buy or rent land for irrigated crop production, they should look for land that has a current water right and

check when that right was allocated and how much water can be used. Water rights stay with the property and are documented on the property's deed.

Water quality. Farmers are responsible for monitoring water quality on their farm to identify health and food concerns; therefore, it is important to have irrigation and drinking water tested. For groundwater, you can learn a lot about a well and its potential for contamination from the well log (created by the well drilling company during drilling), which describes the soil and rock found, depth to first water, and other characteristics. Water-quality tests commonly check for contaminants like arsenic or other heavy metals, nitrate/nitrogen, bacteria, and particular pesticides (if contamination is suspected).

Water quantity. As a farmer and landowner, you must think about the extremes: how to cope with excess water, and how to manage periods of drought. Is your farm prone to flooding? Do you have a plan for animals or crops during a summer drought? To address these problems, you may need to establish diversion ditches, artificial drainage (such as tile drains for excess water), or a water catchment for storage for drought. You may need permits for constructing these types of improvements, so check with your state's water resources department or county land use commission before starting any project.

Climate

Climate is the predictable pattern of temperature and rainfall across the seasons. Sun exposure, rainfall, weather patterns, air movement, and frost all contribute to the climate in your region. All crops need a minimum number of heat units (degree-days) over time to mature and ripen. Certain tree crops also have a chilling requirement. Decisions about what crops to grow should be based on accurate climate data for your location. Key characteristics to investigate include annual precipitation, length of growing season, average high and low temperatures throughout the year, first and last frost dates, and number of frost-free days. Hardiness zones are determined by the average low temperature on the coldest night of the year. This designation predicts the ability of crop plants to overwinter in your location. The USDA Plant Hardiness Zone Map (see Resources) can help you determine which crop plants are likely to thrive in your region. For more detailed location-specific information, check with your state Cooperative Extension Service.

Microclimate. A microclimate is a localized climate effect unique to a small area and different from the larger area surrounding it. Examples of microclimates include:

- Places where air drains and collects on the land
- Changes in wind velocity and wind patterns as a result of physical or natural features
- Temperature and humidity moderation due to the presence of a body of water
- Physical features that affect solar radiation in the area
- Presence or absence of frosts in contrast to surrounding areas

A microclimate can be very small — such as a garden area that is warmer due to heat reflecting from an adjacent building — or the affected area can be larger, such as a valley where cool air collects at night creating an area that is more prone to frost.

In wrapping up this discussion of climate, think through the following key questions:

- What is the climate in your area? Refer to what you discovered from the USDA Plant Hardiness Zone Map.
- Do the crops you want to grow thrive in this climate? Consider the key events in the development of your crop (such as germination, flowering, pollination, and harvest).
- What kind of microclimates exist on your farm that would present problems or opportunities for growing particular crops?
- Can you modify the growing conditions by planting windbreaks or installing a greenhouse or high tunnel, for example?

Location

The location of your farm also has a major influence on how you market what you grow. We have already covered some important location-related variables such as climate, soils, and water. Factors that influence marketing include proximity to population centers, transportation infrastructure, market accessibility, and distance to processing facilities. In addition, local ordinances may encourage or restrict certain activities. If you want to develop an agritourism enterprise, for

PRODUCTION GOALS AND CONSIDERATIONS

PRODUCTION GOAL	LOCATION CONSIDERATION
Growing open-pollinated seed crops	Isolated fields far away from other related crops
Developing agritourism	Appropriate zoning laws
Growing blueberries	Access to adequate irrigation water
Direct marketing products (farmers' market, community-supported agriculture)	Proximity to town or urban area
Sharing farm equipment	Neighboring farmers open to working together
Having farm stand	Safe, easy access road for customers

example, check with your local government to see if there are limitations and what variances are allowed before moving forward with your plan.

The Production Goals and Considerations table on the opposite page lists a number of production goals and the associated location considerations that might need to be taken into account to meet that goal.

BUILDING THE WHOLE FARM PLAN

There are two final activities to integrate what you have learned from this chapter into your whole farm plan.

Create a Farm Map

A farm map provides a visual way to inventory your farm, identify and understand its physical resources, determine land limitations, and consider enterprise options. If your map is to scale, it can help you accurately plan your farm layout and evaluate the efficiency of certain management tasks as you walk or drive around your farm.

You can use aerial photos to begin the mapping process. Local soil and water conservation district offices have high-quality aerial photos that you can obtain. You can also print an aerial photo off the Internet using Google Earth or the Web Soil Survey. Or you can use the soil map you created earlier in this chapter.

The images below show two examples of farm maps, one created from an aerial photo, the other hand drawn.

SAMPLE FARM MAPS

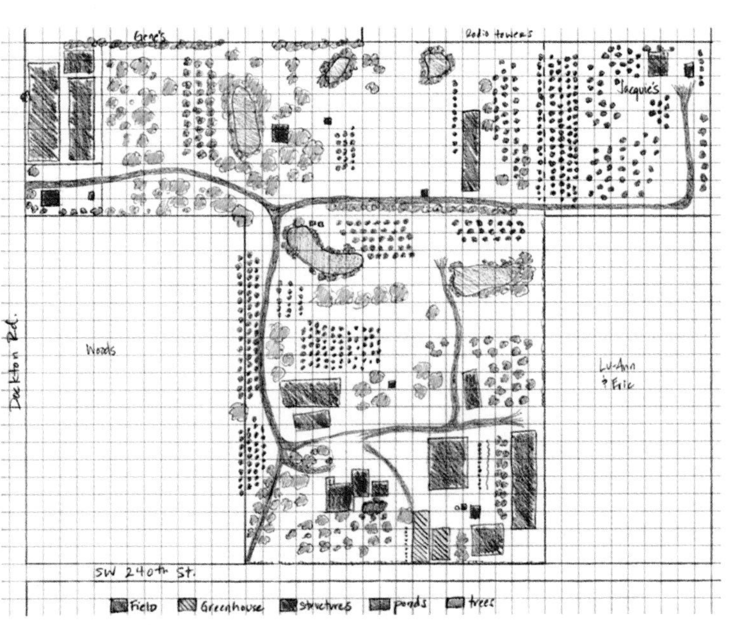

A farm map provides a visual way to identify the farm's physical resources and plan the layout of fields and infrastructure. A farm map may be created from an aerial photo (left) or be hand drawn (right).

FARM PLAN

FARM MAP

Now it's your turn. A farmstead map, no matter how simple, is helpful in the planning process. There is no prepared worksheet for this activity. To create your map, first download an aerial map of your property from the Web or simply sketch a map on paper. Then identify existing natural features and infrastructure on the property. Examples include:

- **Natural features:** Landforms, field boundaries, soil types, slope or other soil limitations, vegetation types, land uses, water
- **Existing infrastructure:** Roads, buildings, fencing, structures, and other major infrastructure (a hoop house, for example)

Then add any new features or infrastructure that you think may be needed and identify them as such. You can create your map by hand or on your computer (if you have those skills). If your map is hand drawn, you can scan it and save it as a PDF or image file. You can then include a printed copy of the map when you assemble your whole farm plan in a binder or folder.

SWOT ANALYSIS

The SWOT Analysis worksheet (appendix 1-13) will help you conduct a SWOT analysis for your farm. Try to fill in at least one or two items for each category. If you are not able to do this now, that's okay. You can revisit this activity as you work through other chapters in the book.

Conduct a SWOT Analysis

A SWOT analysis — a list of strengths, weaknesses, opportunities, and threats — is another useful planning tool for identifying your advantages (personal, farm related, financial, and others) and the obstacles you may face in trying to fulfill your business goals and plans. The process of completing a SWOT analysis helps you better understand what you can control, where your risks are, where improvement is needed, and what direction is best for your business. The farm map you created provides important farm-level information in terms of soil, water, microclimate, location, and more. The Types of SWOT Variables table on the following page shows some of the types of things to consider in each category. Note that strengths and weaknesses are *internal factors*, while opportunities and threats are *external factors*.

TYPES OF SWOT VARIABLES

STRENGTHS

- Knowledge and experience of farm team
- Good infrastructure
- Access to irrigation water
- Land certified organic
- Equipment inventory
- Healthy finances

WEAKNESSES

- Natural limitations such as climate, soil type, and water access
- Lacking or dilapidated infrastructure and equipment
- Relational problems within the farm team
- Lack of experience within the farm team

OPPORTUNITIES

- Consumer demand, market niches, and access to markets
- Local connections
- Education programs and resources

THREATS

- Rising input costs
- Lack of access to labor
- Competition from other farms in the area

PUTTING IT ALL TOGETHER

No matter where you are in the process of building your business, a number of key steps can help bring your dreams to fruition:

- Consider all the people involved in developing your farm and how you can work together to define and reach your goals.
- Assess the natural resources that are the basis of your farm — the opportunities and limitations they present, and how to manage them wisely.
- Work on a planning process that helps you identify your vision, mission, goals, and strategies.
- Create a whole farm plan that describes your farm and the direction you want to go, and think of this plan as something to continually work on and revise.

GOODFOOT FARM

ENHANCING YOUR QUALITY OF LIFE

Beth Hoinacki and her husband, Adam, manage Goodfoot Farm, a 10-acre farm located in Kings Valley near Philomath, Oregon. Beth and her husband have owned the farm since 1999.

"When we started looking for a farm, we didn't really have an idea of what that would look like. The only parameter I had well defined was that I wanted it to be far enough outside of town to have what I considered a rural identity and at the same time be close enough to town to access markets," Beth says.

Their current property was the first place they looked at. It had the basics of what they felt they needed to farm: decent soils and water rights. The farm is on a north-facing slope, which is less than ideal, and in a cold microclimate. But they have come to terms with those drawbacks. Beginning with the two acres of blueberries that were already on the farm when they purchased it, Beth and Adam have expanded the operation to include a variety of vegetable crops, including a number of fall crops that can store through the winter months.

In 2009, Beth took the Growing Farms workshop series through Oregon State University. "One of the big take-home messages for me from that class was matching my production system with the potential of my farm, and also with my personality. So my personal assets and my husband's personal assets needed to match what

we wanted to grow and how we wanted to sell it. Originally we didn't foresee ourselves as being your typical diversified market farm. And then I realized that I had a market farmer inside me and I needed to let her out." Initially, they sold primarily through the First Alternative Cooperative grocery store in Corvallis. More recently, they have started marketing at the Corvallis farmers' market and also offer a CSA program.

Beth notes several important values that have guided their decisions: "One of them is to have a community element, to have our markets be local. Another was to be organic. We're certified organic and we're Demeter Biodynamic certified, and that's always been important to us in developing the farm — that we keep things in a natural balance."

Another value that has guided the development of their farm business is balance in their lifestyle. "We're not willing to be dirt-poor, which is why we're somewhat committed to maintaining off-farm jobs. We can live simply, but being realistic about how we want to live and how farming fits into that is really important as we move forward. So we take the time to reevaluate what's needed to make it work for the family.

"Everyone asks me where the name Goodfoot Farm came from," remarks Beth. "The fact is there is no real story about where the name comes from. Simply put, it captures what our farm means to us and how we want to farm. Getting on the 'good foot' means you have an idea about what you want to be doing, that you're going to get something done, in the right way, and that you have a positive attitude about it."

SLOW HAND FARM

DO WHAT YOU ENJOY

Slow Hand Farm

OREGON

Josh Volk is a farmer, consultant, and entrepreneur. He was inspired to get into farming through his experience in an urban community-garden program.

"I was working as a mechanical engineer in Silicon Valley and volunteering in a community garden in East Palo Alto," says Josh. "It was a neighborhood that had been totally left behind by all the wealth and development in the surrounding communities, and this community garden was the one place where kids could come and play."

Inspired by his experience with the community garden, Josh left his engineering job and, for the next 15 years, worked on a number of vegetable farms learning production techniques. With that farming experience under his belt, Josh began circling back to his community-garden roots. "I didn't have any money, and I didn't have any land when I started out," remarks Josh. "So I looked for somebody who had some land that they wanted to lease, and with water because irrigation was also necessary for what I was planning to do."

From 2009 to 2013, he managed Slow Hand Farm, a small, year-round vegetable CSA on Sauvie Island, just outside of Portland, Oregon.

"I really wanted to start a farm that had a lot of components that I had initially been interested in," states Josh. "In particular, a farm where I could work the soil by hand, without tractors or machinery." According to Josh, the name Slow Hand Farm integrates concepts from the slow food movement and hand-scale (nonmechanized) farming.

His commitment to appropriate technology also extended to CSA delivery. With just about a quarter of an acre in production, Josh was able to deliver shares by bike to members in various locations throughout Portland. Another key aspect of Slow Hand Farm is that it was intentionally managed as a part-time farm, with Josh working only on Mondays and Thursdays. "It was a pretty simple farm, especially in terms of the number of crops. I probably grew 30 or 40 different crops — much less than other CSAs I have worked on," Josh says. "That's the way that I wanted to farm, and that's the way that I enjoyed it."

In 2013, Josh folded Slow Hand Farm's CSA into the startup of a cooperative farm, Our Table, in Sherwood, Oregon. He has made his way back to the urban setting and is now growing with Cully Neighborhood Farm in Portland, Oregon, as well as continuing to do consulting and workshops for other farmers. Slow Hand Farm is profiled along with 14 other farms in Josh's book *Compact Farms: 15 Proven Plans for Market Farms on 5 Acres or Less.*

DO IT

Blue Fox Farm crew Ellyn Greene, Courtney Arts, and Adeline DeMichiel finish planting beds.

CHAPTER TWO

FARM INFRASTRUCTURE, LABOR, AND ENERGY

This chapter encourages you to think through the key operational components of your farm. To use a computer analogy, this chapter is about both the hardware (equipment and infrastructure) and software (people and processes) that let you reach your business goals. The topics addressed in "Do It" connect with all the other chapters in this book and help you develop a holistic view of your farm operations. For example:

In chapter 1, "Dream It," you evaluated the potential for irrigating your property. In this chapter, we take that topic a step further and look at the infrastructure needed to actually deliver water to crops or animals.

In "Do It" we highlight several alternative energy options for your farm. In chapter 3, "Sell It," you will learn how to use such energy conservation measures to create an identity for your farm that can be used to connect with customers in advertising and outreach materials.

THE GOALS OF THIS CHAPTER ARE TO HELP YOU:

- Understand the decision-making process for investments in equipment, buildings, fences, and other farm infrastructure.
- Develop a preliminary list of season-extension and irrigation options that may be appropriate for your farm.
- Know the basic requirements of hiring labor and be able to apply the laws and regulations associated with farm labor to your own business.
- Analyze the energy needs of your farm and explore possible energy-efficiency projects for the farm.

EQUIPMENT AND INFRASTRUCTURE

Think of equipment and infrastructure as the basic framework of your farm or ranch. They support your enterprise and facilitate its growth and success. To get started building this framework, it is important to assess what you have, what you need, and how you can acquire what's needed.

What is the difference between infrastructure and equipment? Infrastructure includes things that are permanently fixed on the farm. Equipment, on the other hand, includes all the tools and movable items that help you do the work of the farm. Listed on the following page are some items that fall into these two categories.

With a basic understanding of equipment and infrastructure, you are ready to inventory the physical resources on your farm. This assessment is an important first step before deciding what items you need to acquire or purchase. Here are a few guidelines to get you started.

If possible, take a full year to observe your farm site before investing a large amount of money in permanent infrastructure. Your observations over the course of four seasons may reveal important information about what is needed and where to locate particular infrastructure components such as fence lines, roads, and outbuildings.

During periods of heavy rain, observe where the land floods or drains more quickly. In winter, make note of areas where the snow melts first. Note wind patterns and effects to determine where a windbreak may be necessary. These are just a few examples of observations that can give you valuable information before planting crops or investing in permanent infrastructure.

Think about your acquisition strategy and weigh the pros and cons of buying used versus buying new. How will you purchase and acquire your equipment? Visit new and used equipment dealers, auctions, and estate sales to get an idea of the costs of purchasing equipment.

Adeline DeMichiel manages the root washer at Blue Fox Farm. The ability to wash and grade root crops offers farms access to more markets.

INFRASTRUCTURE RESOURCES

Greenhouses
Barns
Coolers
Dryers and dry storage
Freezers
Hoop houses
Packing shed
Fences and gates
Storage and/or tool sheds
Irrigation pump and main line
Well and pumphouse
Animal housing and enclosures
Livestock troughs and feeders
Slaughter/processing facilities
Certified kitchen
Solar panels
Pond
Field drainage systems
Roads
Farmhouse
Office
Farm stand
Employee housing

EQUIPMENT RESOURCES

Tractors
Trucks
Trailers
ATV
Tractor implements
Harvesting equipment
Hand tools
Pruning ladders
Power tools
Irrigation pipe and hand line
Driplines
Row covers
Packing equipment
Movable broiler pens
Mobile chicken coop
Electrified fencing components
Processing equipment for value-added products
Milking equipment
Produce-washing equipment
Product-packaging equipment
Marketing equipment such as a canopy and display tables and racks

Planning and Purchasing

What equipment and infrastructure are needed on your farm? As you work through your inventory and assessment and develop your priority list, take your time and carefully weigh the costs and benefits. Acquiring equipment and building infrastructure are some of the more costly endeavors for a new farmer, and it can be frustrating if you do not have a clear budget or plan of action. Here are a few suggestions as you get started:

- **Think big, start small.** Purchase only exactly what you need and do not buy equipment just for the sake of buying equipment.
- **Consider the appropriate scale for your equipment purchases.** How much horsepower is needed? What type of implements should the tractor be able to handle? Think toward future needs if your intention is to expand and become a larger operation.
- **Plan your acquisition strategy.** How will you purchase needed equipment and infrastructure? Be specific about whether you will be seeking a bank loan, arranging for credit financing with the dealer, paying cash, or borrowing from a friend or family member.
- **Check brush patches** and other outlying areas of the farm for old, discarded implements such as harrows and discs that could be restored.
- **Do your research** and shop around for the best deals.
- **Visit other farms** and talk to other farmers about what you might need for your enterprise. Ask them how they acquired their equipment. Experienced farmers can help you avoid costly mistakes.
- **Consider renting or trying out equipment** before you purchase it to really understand what fits your farm and your enterprise.
- **Look for opportunities to share equipment** with other farmers, or contract out for some activities.

Deciding on a Tractor That Fits Your Needs

Purchasing a tractor is often at the top of the list of needs for a new farm business. Listed below are some of the most important purchase considerations.

Servicing. Before purchasing a particular make and model of tractor, find out if you can get parts for it. Also, think about who will do the servicing. Can you do it yourself, or will you need the help of a mechanic? These questions apply to new tractors as well as used, since tractor dealers can go out of business. Tractor implements can break down. You may need a welder to do repairs or a machinist to fabricate specialty parts for you.

Size and horsepower. Think about what you need your tractor to do and how much power you need it to have. It is most cost effective to purchase a tractor that is ideal for your everyday uses, not for the major projects that occur only once every few years; you can always rent a tractor for the big projects. There are three different types of horsepower to consider: engine, drawbar, and PTO horsepower. Evaluate all three of these to get the most accurate assessment of the tractor's capabilities.

Power takeoff (PTO). The PTO is connected to the engine or differential and allows the tractor to run powered implements such as mowers and rotary tillers. PTO horsepower refers to the percentage of the engine horsepower that the shaft transfers to the implement. There are two types of PTO: standard and live. Standard PTO only works when the clutch pedal is released; live PTO runs regardless of whether the clutch is engaged or not. In used tractors with a standard PTO, make sure the PTO engages and comes to a full stop when disengaged. If the PTO continues turning, even slightly, this is a sign that it is worn and will be an expensive repair. Not all tractors have a PTO, so you will need to decide if this is an essential feature for your system.

Tires. Tires have different ratings that correspond to different types of traction in the field.

New or used? There are pluses and minuses either way. A new tractor is more expensive, but it will most likely come with a warranty and you will have access to parts and service. You can also finance new tractors through banks or the tractor dealer. Buying a used tractor can save thousands of dollars in the purchase price, but it may have more problems over the long run that can add significantly to the cost. How well does the tractor run? What was it used for in the past? Consider getting the opinion of an expert mechanic before you commit.

Specialty tractors are often available only as used models. For example, vegetable farms often use older rear-engine or offset cultivating tractors (Allis-Chalmers G and Farmall Cub are two examples) as they have an open frame that allows for good visibility while weeding plants. These older tractors can be hard to locate and they usually don't have PTO.

Implements. Does your tractor come with a loader? What kind of implements can the tractor pull and can you purchase what you need from the same tractor dealer?

Other factors. Also consider tire spacing/axle length, which determines what row spacing you can use in the field; wheel width on some tractors is adjustable. Consider tractor weight, stability, and maneuverability — analyze the optimum combination of these factors for your situation. Will you need a four-wheel-drive tractor? Would a walk-behind tractor be appropriate for your size operation?

You'll find detailed examples of tractor assessments in an annual cropping system in appendix 2-1, in a grass-based livestock system in appendix 2-2, and in a perennial cropping system in appendix 2-3. Each assessment includes a hypothetical analysis of what a farmer or rancher might be looking for in a tractor. These are presented to give you an idea of what an assessment might look like for that type of system; every farm has different needs, so assessments will vary accordingly.

Now it's your turn. After studying the tractor examples above, turn to appendix 2-4 and fill in the blank table for your own farm. Fill out the table as best you can, then save it to refer to when you begin completing the Equipment and Infrastructure Assessment worksheet (appendix 1-14). As you work through your own tractor needs assessment, note how what you produce on your farm (your choice of crop or livestock system) influences your equipment needs and options.

Farmers Share Their Experience

ESSENTIAL EQUIPMENT AND INFRASTRUCTURE

SEVERAL FARMERS OFFER THEIR PERSPECTIVE BELOW on the equipment and infrastructure that they feel are key to the success of their farm business. Note how each farmer has a different list of essential equipment, depending on the size of the farm, the type of crop or livestock product, the location of the farm and related climate variables such as temperature and rainfall, the specific system each farmer has developed, and how he or she markets the product. After reading about other farmers' experiences, make a preliminary list of all the infrastructure and equipment that you currently have on your farm. Save your list for the formal assessment activity that you will work on later in this chapter (page 75).

SLOW HAND FARM

JOSH VOLK, SLOW HAND FARM (profile on page 46)

In the first season, we started out packing CSA shares in the shade of a couple of fruit trees. We hauled our tools back and forth to the farm every day that we were here, so there was basically no infrastructure on the farm other than the water that was supplied to the field for us. Any starts that we grew that year were grown off-site on different properties, so everything was being brought here.

After that first year, the owner put in a hoop house, which we were able to use for propagation. And it had electric power, so we were able to use electric heat mats. We used a battery timer to water those starts every day. He also gave us access to a corner of one of his barns to store our washing and packing supplies, which were basically a simple table and washtub and some sawhorses with pieces of lumber on which we hang bags. In another little corner of another barn, he let us store our tools, so we didn't have to haul them back and forth.

I primarily use hand tools. The only power tool that I use with any kind of frequency is an electric battery-powered lawnmower. And that's just for maintaining the edges. Hand tools, relative to power tools, are pretty inexpensive, but I get really high-quality tools. And I have a couple of sources that I like, so I generally spend a couple hundred dollars on tools in a season.

RENARD TURNER, VANGUARD RANCH (profile on page 132)

We have about 60 acres that are fenced and cross fenced for the goats. The basic infrastructure is a lot of fencing, automatic waterers, livestock-guardian dogs, and run-in shelters — that's really all the goats need. Our climate's pretty temperate, but in the wintertime, they'll spend time in the barn. They're always free-range, so they come and go as they please. A key to the success of my fences is that my goats don't jump fences and climb because they have the myotonic genetics — that makes them so much easier to maintain. I had Kikos, and they would jump my 4½-foot fence. These goats never do that.

BETH HOINACKI, GOODFOOT FARM (profile on page 44)

We have a BCS two-wheeled, walk-behind tractor. It has a tiller implement, a potato digger, a furrower, and a brush hog. For the small scale of our operation, it's my primary tool out in the field during the season. That tool really works well.

I'd say for any farmer, you want to match your equipment to the scale of your operation. The bigger the tractor, the more headland space you need to turn around. We started out with 12-foot headland space and we just couldn't turn the tractor around to come back down the other row so we had to increase that space to 16 feet. If we were just using the walk-behind, I could turn around in 8 feet of tractor space. More headland space means less space for production, and if you don't have a lot of land you should consider this when selecting a tractor.

CARLA GREEN, SWEET HOME FARMS (profile on page 216)

When we purchased this place, it had what we thought was a really great fence — a five-wire New Zealand fence on the perimeter. It was one of the big draws. It didn't have much interior fencing, but the perimeter was really good.

It had this nice old barn, but it did not have a very good barnyard. In the area just to the south of here, there was about 2½ feet of muck the first winter. So we began to realize very quickly that we needed to manage mud.

And we also realized, too, that we needed a better, bigger barn. As we gained more and more animals on the property and struggled to keep that New Zealand fence hot, we also got increasingly concerned about having animals out on the road. So in the end, we had to put in a huge amount of fencing. Fencing has been critical.

We also built a large barn for feeding and hay and equipment storage, and more recently we built a compost barn to manage the bedding.

CHRIS JAGGER, BLUE FOX FARM (profile on page 168)

One of the key pieces of equipment that we cannot live without, starting in the propagation greenhouse, is the vacuum seeder. Once we got it, we thought: What were we thinking all these years when we didn't have it?

Another important thing was a big enough tractor. We worked with an underpowered tractor for a long time. Most of our work is now done with a 68-horsepower tractor that's four-wheel drive. We couldn't live without four-wheel drive. That's huge.

And then we have all the standard pieces of equipment that most people utilize that I guess we wouldn't live without — a disc, a moldboard plow, and the rototiller that we do all of our seedbed prep with.

The walk-in cooler makes a huge difference in terms of shelf life and storability. As does a refrigerated truck. Then there's our root washer, which we built a couple of years ago.

Dry storage inside is also important. We expanded our barn and made it a completely dry facility with drying shelves. We used to cure all of our onions outside and prayed for no rain. The inside storage has been huge for us.

SARAHLEE LAWRENCE, RAINSHADOW ORGANICS (profile on page 218)

My transition to small-scale agriculture has been an adventure. I grew up on a farm that was big tractors and harrow beds and big hay barns, semis, and equipment — a one-man show with big equipment.

Pulling back into a small-scale operation where I had to learn how to use my own hands and use really great hand tools and small implements has been a process for sure. My first season, I needed to have a deer fence since we have a lot of wildlife here, including elk. I needed a seven-foot fence that could be eight feet. That was a big part of my first year.

I also had to have a greenhouse if I wanted to grow anything that wasn't super-hardy. Going into my second year, I learned that I needed refrigerated space. It's extremely hot and dry here, and things wilt immediately.

The second thing that I had to have going into my second year was a wash station. Before that, I washed everything on the ground, squatted over, so I needed to improve the ergonomics of my work.

Going into my third year, I built a lot of hoop houses. I'm really trying to expand what we do. We also started our meat operation in our third year. We built a lot of fencing, pig houses, chicken houses, mobile chicken tractors. In our third year, all of our infrastructure money pretty much went toward animal infrastructure.

SEASON-EXTENSION EQUIPMENT FOR VEGETABLES AND BERRIES

For many farmers, season-extension equipment is another important item on their list of must-haves. Given the range of techniques and potential costs involved, we address season extension here, with a particular emphasis on the infrastructure and equipment needed for each option.

Many farmers are using season-extension techniques to gain a competitive advantage in the marketplace, generate cash flow at slower times of the year, and enhance overall profitability. Using hoop houses or row covers, for example, can provide a head start in late winter or early spring, resulting in earlier harvest of crops such as cucumbers, tomatoes, basil, peppers, and berries. At the other end of the growing season, similar techniques can keep fall vegetable crops such as brassicas, lettuce, and tomatoes in production into winter. In either case, farmers can charge a premium price by offering a local product at a time of lower supply. Season-extension strategies thus become a way of managing risk by optimizing production and farm income.

Advantages and Disadvantages of Season Extension

Understanding the basic advantages and disadvantages of season extension will help you decide which strategy or technique is most appropriate for your particular farm. The lists below highlight some of the major considerations.

If you want to extend the production season on your farm, you have two basic options: you can grow overwintering crops and varieties that are adapted to colder weather and shorter days, or you can modify the environment to

ADVANTAGES

Season extension results in a longer period of time with cash flow/income.

A market advantage can be gained by offering product earlier or later in the season than other farmers, allowing you to charge a higher price.

Extending the growing season maintains a relationship with customers over a longer period of time.

Season extension allows for greater diversity in what you produce and market.

You can retain employees through an extended growing season.

Season extension helps you manage risk related to weather extremes.

DISADVANTAGES

Season extension is management and labor intensive.

Season extension requires different expertise than standard field production.

With season extension, there is a learning curve that will take a while to work through.

Season-extension practices are costly and require an investment in materials and maintenance.

Depending on labor availability, farmers should consider that a longer growing season means a longer period of work without a break or vacation.

Some season-extension strategies use materials that are dependent on fossil fuels for manufacturing (e.g., polyethylene).

manage soil moisture and increase the soil and air temperature in which the crops are grown. We address each of these options in turn.

Choice of Crops/Varieties

Choosing crops and varieties adapted to cooler temperatures is the simplest approach for extending the growing season. Farmers can look for cold-hardy, early-maturing crops that can produce earlier in spring or that can be planted in fall and survive temperatures ranging between 20 and 40°F (–7 to 4°C). The table on the right lists several crops that are adapted to fall and winter growing conditions.

The minimum temperatures shown in the table indicate the lowest temperature at which the crop can survive. If the crop can tolerate some damage, it may be able to survive even lower temperatures. While these crops can survive cold temperatures, they do not grow much during cold winter months. Some are planted in mid- to late summer so they mature and store in the field over the winter. Others are established in fall so they can get an early start for a late-spring or early-summer harvest. Other cold-adapted crops to consider include garlic, onions, endive, escarole, kohlrabi, parsley, asparagus, rhubarb, rutabaga, peas, wheat, barley, oats, canola, flax, and fava beans.

EXAMPLES OF CROPS ADAPTED TO FALL/WINTER GROWING CONDITIONS

VERY HARDY	HARDY	SEMI-HARDY
Leeks (0°F/-18°C)	Cabbage (10°F/-12°C)	Lettuce (15°F/-9°C)
Parsnip roots (0°F/-18°C)	Broccoli (10°F/-12°C)	Carrot — roots (15°F/-9°C)
Spinach (0°F/-18°C)	Brussels sprouts (10°F/-12°C)	Beets, chard (15°F/-9°C)
Kale (0°F/-18°C)	Cauliflower (10°F/-12°C)	Mustard (20°F/-7°C)

Source: *Fall and Winter Vegetable Gardening in the Pacific Northwest*. Pacific Northwest Extension Publication PNW 548, 2001.

Environment-Modification Techniques

There are a number of techniques for modifying soil and air temperatures, all aimed at creating a more favorable environment for crop growth. There are four main options: row covers, plastic mulches, high tunnels, and multibay high tunnels.

ROW COVERS

Row covers are movable barriers that can be laid over crop beds at certain times in the season, usually for frost protection or insect control. Floating row covers lie directly over the crop and may cover multiple rows; hoop covers (low tunnels) usually cover just a single bed. There are various types of row cover materials — from clear polyethylene-type plastic to mesh fabric — that allow for varying degrees of frost protection.

Row covers cost much less than high tunnels and can be an inexpensive way to extend your season. Costs vary considerably depending on the specific product, so shop around for the best row cover prices. Potential concerns when using row covers include the following:

- The cover must be secured on windy days.
- Additional frost protection may be needed.
- Row covers often need support to keep the cover from damaging crop plants.
- Weed control can be a challenge.

To learn more about row covers, watch the video *How to Install Row Covers on Vegetable Crops* from the University of Minnesota (see Resources).

PLASTIC MULCHES

A plastic mulch usually consists of some type of polyethylene material that is used to cover the soil only. Transplants are usually planted through the plastic sheets. A plastic mulch helps with weed and insect control, protects planting beds from erosion, promotes warming of the soil for better crop growth and yields, and can help prevent soil saturation during wet weather. There are several types of mulches — black, clear, white, green infrared-transmitting (IRT) plastic, and biodegradable.

Plastic mulch materials vary in cost, and each type has different benefits and challenges. Key considerations when using plastic mulches include the following:

- The effect of the mulch on soil warming is determined mainly by the color of the mulch and how closely the mulch is in contact with the soil.

ROW COVER

PLASTIC MULCH

- There are a variety of techniques and machines for laying the plastic sheeting.
- The plastic should be well anchored in windy areas.
- Biodegradability and disposal of plastic mulch material can create problems.

To learn more about one farm's approach to using plastic mulches, watch the video *Laying Plastic Mulch at Rooster Organics* presented by Abundant Harvest Organics (see Resources).

HIGH TUNNELS

A high tunnel is a semipermanent structure in the field, covering multiple beds, that allows room for people to stand, work, and operate equipment. High tunnels usually have a framework of galvanized steel or PVC pipe covered with durable, translucent polyethylene material that has been treated to protect against damage from UV light. This creates a solar-heated greenhouse that protects crops from adverse weather damage and creates a more ideal growing environment. Some growers have developed movable high tunnels that can be relocated to different sections of the farm.

The costs of a high tunnel (plastic and frame structure) can vary widely depending on size of the tunnel, materials, construction methods, and vendor. There are additional costs for site preparation, special end-wall construction, water lines, and other accessories (such as electricity, fans, and lighting). Key considerations with high tunnels include:

- Proper spacing of hoop structures
- Anchoring the framework for high winds
- Cost and maintenance of built-in features for ventilation and temperature control
- Removal of ice and snow

To learn more about high tunnels, watch the video *High Tunnel Construction* from the University of Kentucky (see Resources).

HIGH TUNNEL

MULTIBAY HIGH TUNNEL

MULTIBAY HIGH TUNNELS

Multibay high tunnels are a type of high tunnel widely used in Europe and recently introduced to the United States. The polytunnel, multibay structures are large enough to drive tractors through. Sizes range from 18 to 30 feet wide and between 10 and 18 feet high. The Haygrove system of multibay high tunnels was developed in the United Kingdom, and there are various sizes and structures on the market; most are temporary structures.

The cost for a Haygrove high tunnel depends on size and type of structure. Investigate farms using these tunnels before purchasing. The primary concerns with the Haygrove system are cost, construction expertise, and securing the tunnels against wind damage.

Farmers Share Their Experience

PROCURING EQUIPMENT AND DEVELOPING INFRASTRUCTURE

INFRASTRUCTURE AND EQUIPMENT are key components of a successful farm business. So far in this chapter, we have described what is included in those categories and given you the opportunity to think about two specific aspects of equipment and infrastructure that are often mentioned by vegetable and small fruit farmers as being essential for their operation: tractors and season extension. These two items (as with many other equipment and infrastructure needs) can be costly, so how do you make wise decisions about farm infrastructure and equipment?

BETH HOINACKI, GOODFOOT FARM

(profile on page 44)

When we purchased the farm, there was basically nothing here. And so we needed money for fencing. We can't grow anything here without a deer fence. We needed money for irrigation systems. We needed money for structures, buildings, barns. We're in the process now of putting up hoop houses. We've integrated animals into the system that require a certain amount of infrastructure.

The idea has always been to generate enough off-farm income to pay for those capital expenses and provide our income until we could get the system to where production would match that infrastructure and would fulfill the potential of that infrastructure.

If you look at what we have, we're very equipment-heavy versus what we're producing. But the idea has been not to depend on production until we had the infrastructure in place to mostly be efficient.

So, for example, we have a Tuff-bilt cultivating tractor. We aren't using it at the moment, but from looking at other farming systems and seeing where our farm is going, we saw that a cultivating tractor was key to working an organic vegetable system efficiently. I had the opportunity to buy one, and I jumped on it because that's where I saw things going. We'll probably put a new engine in it and get it up and running next year, so we'll be able to manage our increased production without so much more labor.

MIMO DAVIS AND MIRANDA DUSCHACK, URBAN BUDS: CITY GROWN FLOWERS

(profile on page 129)

Mimo: We bought the farm because of the glass greenhouse. That was really it — the old,

behemoth 1950 Lord and Burnham glass greenhouse. And it was in pretty bad shape.

It was four city lots but sold as one unit — two of which had never had buildings on it. So we could plow, which was very unusual for urban farms. When people talk about urban farming on vacant land right in the middle of the city, it's all rubble, so you have to bring in soil to create permanent raised beds on top of the fill dirt. The investment is phenomenal.

Miranda: Now we have the glass greenhouse, the glass propagation house, a plastic greenhouse, a high tunnel, and a hoop house. And then we have various movable caterpillar tunnels, season-extension tools, and what we call field space. In that field space, we have raised beds we made with trucked-in soil and raised beds we made with the soil that was here.

CHRIS JAGGER AND MELANIE KUEGLER, BLUE FOX FARM

(profile on page 168)

Chris: When we first started, we generally bought used equipment because we didn't have the cash flow to support new equipment. As our business became more established and we had less infrastructure to invest in, we had more cash flow to commit to new equipment.

Buying new equipment has been our biggest change recently. It sometimes feels weird to us, but at the same time, we bought a brand-new tractor five years ago and haven't done anything other than regular maintenance. So that's really paid for itself.

Melanie: We have felt more comfortable buying higher-priced newer equipment now because our markets have solidified. In the first few years, you're not really sure what your year is going to look like in terms of income. Now we have a much more solid picture of that.

RANDY KIYOKAWA, KIYOKAWA FAMILY ORCHARDS (profile on page 94)

We didn't do it all at once. It's like how I used to mow the turf rows; I used a riding lawnmower until we could say, 'Okay, I have enough money to buy a regular small tractor.' So, what you see now is very different from what I started with.

SARAHLEE LAWRENCE, RAINSHADOW ORGANICS

(profile on page 218)

My neighbors, who are all conventional farmers who have been farming forever, and mostly older fellows, have been incredibly supportive of me and my efforts. They've helped me get going, make some transitions, and at least get by. I really appreciate them. I have one neighbor in particular who has come through on multiple occasions. This year I got a potato digger from them — this ancient single-row potato digger that we refurbished and I was then able to use.

Key Season-Extension Points

In summary, season extension offers crop farmers a great opportunity to increase production early and late in the growing season and to potentially improve the farm's overall bottom line. To achieve success, however, growers should consider the following key points:

- **Crop selection:** Know which varieties and cultivars are best for winter and early-spring production or are most suited for the modified environment under tunnels or row covers.
- **Site selection:** Think about sun exposure throughout the year and shade from trees or topographical features. Avoid frost pockets and windy sites. In placing greenhouses, pay attention to the orientation of the building relative to the path of the sun. At northern latitudes, an east–west orientation along the length of the greenhouse is better at catching maximum light during winter.
- **Time of planting:** Because season extension shifts the growing season, it is important to know how to adapt seeding and planting times for the type of season-extension technique being used.
- **Growing the crop:** Modify production practices accordingly to provide the best conditions possible for plant growth and reproduction. Understand how your particular season-extension system works so that you can properly control the flow of heat in the soil and in the air to create the optimum environment.
- **Do your research and start small:** Explore online, read articles, and talk to farmers and other experts who have used season-extension techniques. Experiment with different crops to see what works best.

IRRIGATION EQUIPMENT AND INFRASTRUCTURE

Water is a cruical production factor for many farmers and another important area of investment in the farm business. In chapter 1, "Dream It," we looked at water resources from a planning perspective. You learned about assessing water resources on your farm and we addressed the importance of water rights in some regions. In this chapter, we focus on the infrastructure and equipment needed to get water to your crops or livestock. Specifically, we address the following questions:

- What are the major types of irrigation systems?
- What do you need to know to design an appropriate-scale irrigation system?
- What water-quality factors need to be taken into account when setting up irrigation systems?
- What are the costs?

To get water to the field and to create enough pipe pressure to activate sprinklers, an irrigation system must be powered either by the force of gravity or by an irrigation pump.

Gravity-flow system. A gravity-flow irrigation system is a low-cost method for providing water to crops and pasture. Such a system consists of an elevated water source connected to a ditch or main pipe that brings water to the field. Depending on the source, volume, and flow rate, the water from the main pipe can be delivered through secondary pipes and applied through furrows or sprinklers. If your pond or irrigation source is at the highest elevation on your farm, take advantage of this opportunity and design a system that relies on gravity to supplement or replace pumping. This can save you money and time in the long run.

Pump-irrigation system. A pump-irrigation system relies on an external energy source to get water to the field. Most irrigation pumps are centrifugal pumps, which use an impeller to spin the water through a casing by means of centrifugal force. Irrigation pumps can be operated either by an electric motor or a fuel-powered engine. Types of centrifugal pumps include:

- End-suction centrifugal pumps
- Submersible pumps
- Turbine and jet pumps
- Booster pumps

For a full description of each of these types of pumps, see *Irrigation Pump Tutorial: Selecting a Pump Type* (see Resources). Regardless of the type of pump, the key factor is to make sure everything matches in your system — pump, source capacity, nozzle heads — and that it fits together. Do not purchase the pump until you have designed your whole irrigation system. This rule applies whether you are developing an irrigation system from scratch or renovating an existing system that came with your land.

Designing an Irrigation System

Designing a good irrigation system requires knowledge and expertise. To make the most of your investment of time and money, we recommend that you consult with an expert who can help you through the process. Whether you do it yourself or work with a consultant, you need to be familiar with several important irrigation terms and concepts as you begin the design process.

Evapotranspiration. Evapotranspiration (ET) is the total water use in a given area as a result of both plant transpiration and evaporation from the soil and plant canopy.

Field capacity. Field capacity is the moisture content of the soil after the "free" water has drained away — usually the soil moisture content one to two days after a soaking rain. The amount of water in a soil at field capacity varies by soil type (see the Soil at Field Capacity diagram on page 66).

Available water capacity. Available water capacity is the portion of soil water that can be taken up readily by the plants' roots. This determines how much and how often you must irrigate to prevent moisture stress.

SOIL AT FIELD CAPACITY

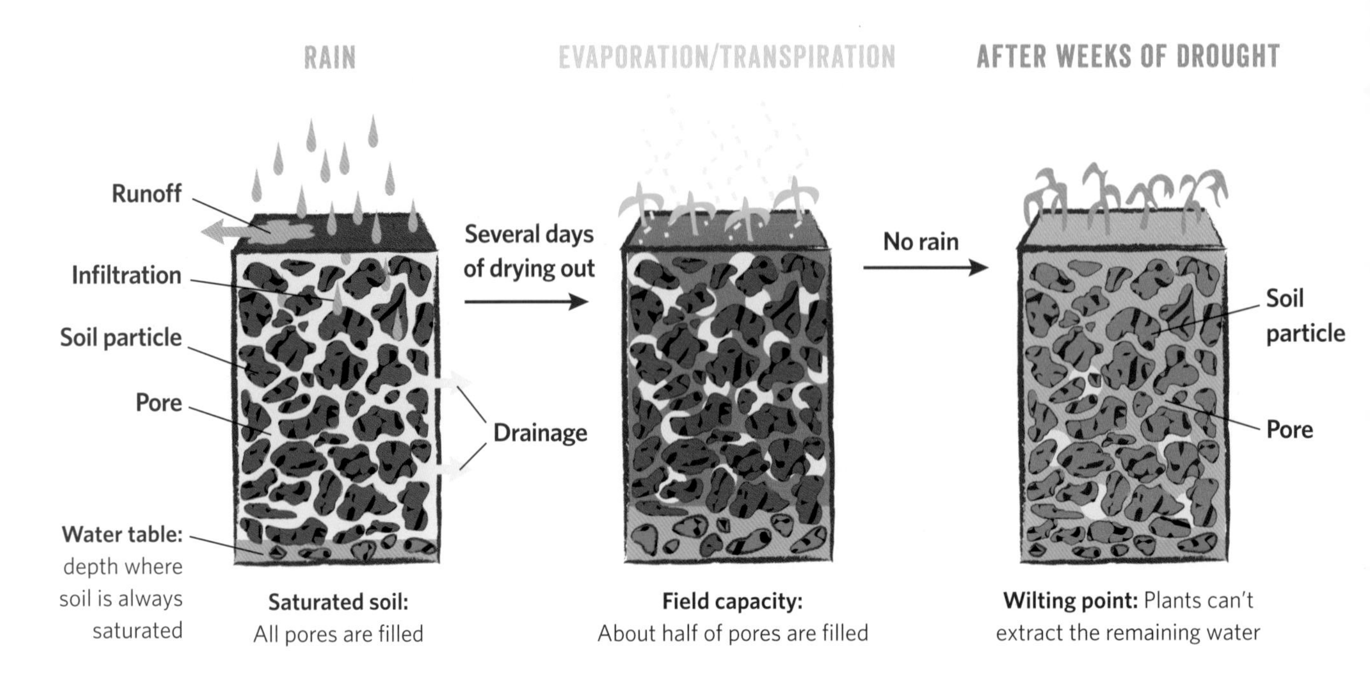

Managed root zone. The root zone varies by crop and growth stage and is influenced by soil depth, compaction layers, and dry soil. The managed root zone is considered to be the upper 75 percent of the root zone, where most plant feeder roots are contained.

Water intake rate. Water intake rate is the speed at which water can enter the soil. Fine-textured soils (such as silt and clay-dominated soil) have slower intake (or infiltration) rates than coarse-textured sandy soils. This influences the maximum application rate.

Maximum application rate. Maximum application rate is the rate at which water can be applied to a given soil without puddling or runoff (how quickly you can apply water). In a farming situation, the application rate is determined by many factors, including soil texture, slope, soil vegetative cover, evapotranspiration rates, and type of irrigation.

To design an irrigation system that is appropriate for your particular crop system, you will need good weather data about your particular location. Nationwide, there are many resources available to help farmers and ranchers monitor evapotranspiration and crop water use. In the Pacific Northwest, farmers can use AgriMet (see Resources), a Web-based resource consisting of agricultural weather stations located throughout the region. Maintained by the Bureau of Reclamation, the stations are located in 70 irrigated agriculture areas. AgriMet estimates evapotranspiration or crop water use for crops grown in the vicinity of its stations. Similar resources are available in other states. Bear in mind that spring and summer rainfall in some regions often varies across the landscape so on-site rain gauges are usually the most accurate way to measure rainfall and are also useful for calibrating sprinklers. Knowing daily crop water use helps farmers schedule irrigation to fit crop needs.

IRRIGATION DESIGN FACTORS

In addition to the basic soil-water concepts summarized above, there are design factors that farmers must consider when designing an irrigation system.

Setting up a main line. For irrigating crops, it is common to lay a larger main line and then have lateral lines coming off that to deliver water to the field or orchard.

Conservation and efficiency of delivery. Water can be lost through inefficient delivery systems, so maximizing efficiency should be a part of your design. Using the most efficient delivery infrastructure, configuring an efficient irrigation layout, matching plant irrigation requirements with systems, and selecting efficient storage systems all contribute to efficient irrigation systems.

Timing and scheduling of irrigation. How and when to deliver water is another consideration. Plant growth, root development,

Set up seasonally, this irrigation system on Blue Fox Farm draws water from a holding pond.

canopy (amount of leaf area), and soil water-holding capacity all play into how and when to water crops. Water use varies by crop and pasture, as well as throughout the season.

Parts and maintenance. Over time, the components of an irrigation system are subject to wear and tear, reducing the performance and efficiency of the system. A poorly functioning irrigation system wastes water, energy, and nutrients that have been applied to the field. Routine maintenance will help keep your irrigation system working at optimum levels. Here are some key tasks to put on your list:

- Replace nozzles every four years at a minimum — in some situations, you may need to replace them more frequently.
- Make sure sprinklers are discharging at the correct angle and impact sprinkler heads are rotating properly. These factors ensure that sprinkler systems will evenly distribute the water.
- Check system operating pressures periodically. If they are below optimal range, check that the pump is working correctly, that the number of sprinklers is not in excess of pump capacity, and that nozzles are not worn.
- Give your irrigation pump an annual checkup. Note the condition of the motor, bearings, electrical connections, suction lines, pipes, and joints. Clean out screens regularly. For a more complete description of pump maintenance and repair procedures, see the ATTRA publication *Maintaining Irrigation Pumps, Motors, and Engines* (see Resources).
- If you are using a portable aluminum main line or lateral, inspect gaskets annually for cracking and/or shrinkage. Remove gaskets during the off-season and store in a dark, cool area such as a garage or barn.
- If possible, store pipe off the ground to discourage rodent nesting, and provide adequate support for the pipe to minimize "bowing" that can otherwise occur during long periods of storage.

Drip irrigation system at Rainshadow Organics conserves water in a desert environment.

Well capacity and peak flow. Well capacity (or specific capacity) is the pumping rate (yield) divided by the water level drawdown in feet. For example, a well with a pumping rate of 120 gallons per minute (gpm) with a drawdown of 6 feet has a specific capacity of 20 gpm per foot of drawdown. Well capacity is used to roughly estimate the maximum yield for the well. A simple way to notice large changes in well performance is to monitor specific capacity. If possible, obtain the specific capacity of the well at the time it was developed and brought online, running the well at the flow rate you expect it to experience in daily use. This will serve as your benchmark for tracking well performance over time. Well capacity declines with the age of the well, usually as a result of plugging and clogging. In designing your irrigation system, the flow requirements (particularly peak flow) for your system should be determined in conjunction with the specific capacity of your water source (well or surface supply).

Water quality. When setting up your system, think about your watershed — where your water comes from and where any excess water goes after it leaves the field. Look for ways to optimize water quality by reducing fertilizer, nitrate, and sediment runoff into streams and rivers. For instance, to allow sufficient water infiltration across a field in furrow irrigation systems, some water will run off the lower end of the field (called tailwater). This runoff can be a problem when it carries sediment, nutrients, or chemical residue into streams. There are systems for managing and reusing tailwater to reduce negative impacts.

The U.S. Clean Water Act controls point source pollution from agriculture by regulating manure management or confined animal feeding operations. It also provides guidance for improving non–point source pollution from agricultural runoff. Your state or county may additionally have best management practices or regulations to protect water resources. For more on agricultural runoff,

read the EPA's *Protecting Water Quality from Agricultural Runoff* (see Resources).

Additionally, practices such as installing fish screens on irrigation intakes or planting riparian buffers along streams can protect fish habitat and may make your enterprise eligible for certification options, such as Salmon-Safe Farm Certification.

On the water delivery side, it is important that you assess the quality of your irrigation water and look for possible negative impacts on crops and livestock. For example, some groundwater sources may be contaminated with arsenic or other elements native to the soil environment. Other wells may have problems with high levels of salts or nitrates in the water. And surface water may contain contaminants that should not be applied to crops that are ready for harvest and consumption. Irrigation water quality can be evaluated with a water-quality test.

A self-flushing irrigation filter reduces irrigation system maintenance at Kiyokawa Family Orchards.

IRRIGATION PUMPS

Your irrigation system design determines the type of irrigation pump and other equipment that you must purchase. You should estimate flow and pressure-head requirements for your system, then select the pump based on those calculations — not on horsepower alone. For existing systems, figure out whether you need a new pump similar in size to the one you are replacing or whether you should adjust capacity/power up or down to suit your needs. In projecting costs for the pump, figure in maintenance costs for the equipment as well as the initial purchase price.

Pressure and head are interchangeable concepts in irrigation. *Head* refers to the height of a vertical column of water. The resulting pressure calculation is based on the fact that a column of water 2.3 feet high is equivalent to 1 pound per square inch (psi) of pressure. So the pressure head at any point where a pressure gauge is located can be converted from pounds per square inch to feet-of-head by multiplying by 2.3. For example, 15 psi is equal to 15 times 2.3, or 34½ feet of head. Most city water systems operate at 50 to 60 psi, which explains why water supply towers are generally about 130 feet above the ground. Flow rate and total dynamic head are determined by the type of irrigation system, the distance from the water source, and the size of the pipes. For a detailed explanation of the calculations involved, see the North Dakota State University Extension publication *Irrigation Water Pumps* (see Resources).

As we indicated at the beginning of this discussion, designing a good irrigation system is a complex endeavor. We recommend you obtain the services of an advisor or consultant who can help you work through the process.

Water Delivery Methods: Crops and Pasture

The final step in your irrigation system design is deciding how to actually deliver the water to the field. Are you going to use some kind of overland flow method, such as furrow irrigation? Or will sprinklers work better for your particular situation? What about drip irrigation? You may be aware of the benefits of drip, but is it appropriate for your situation? There are a number of options to choose from. To help you decide, consider the following key questions.

What are your farm goals and vision? Are there priorities in the mission, vision, and goals for the farm that might guide you? For example, conservation may be very important, so a system that promotes water use efficiency and conservation might be the way to go.

What crops are you growing or do you want to grow? What kind of livestock are you raising? The type of crop or livestock system is another factor that determines the type of irrigation/watering system needed. Certain crops produce better with a particular type of irrigation. Tomatoes, for example, can be particularly productive under drip irrigation. Pasture or hay crops do well with gun-type sprinklers. With wider and deeper rooting zones, tree fruits do well with overhead sprinkler irrigation or microsprinkler irrigation.

What soil types do you have on your farm or ranch? Soil type, particularly soil texture, influences what type of irrigation system is appropriate and how it should be designed. In coarse-textured or sandy soils, it would be difficult (and very inefficient) to get water all the way to the end of a long furrow, so sprinkler irrigation may be the best choice under those conditions. At the other end of the spectrum, in fine-textured silt or clay soils, drip irrigation might be a good choice since it can be regulated to deliver water at a rate that matches the soil's slower infiltration rate.

What labor do you have available to work on irrigating? Different types of irrigation systems have different labor requirements. Systems that require a lot of moving of pipe, or close monitoring of water delivery, may not be the best choice in a solo operation. A related factor is the level of skill and experience within the farm team in managing certain types of irrigation systems.

After thinking through these questions, take the time to learn as much as you can about your different options. Here we include a brief summary of the main types of irrigation methods.

Hand watering. Some hand watering can be a viable option for small-scale operations. Even for farms that use more extensive irrigation methods in the field, starts produced in the greenhouse often require hand watering with hoses and special spray nozzles. For this type of irrigation, consider water pressure, hose type and size, and sprinkler attachments. Once you get past a certain size operation, hand watering is not an effective option.

Components: Hoses, spray nozzles
Setup costs: Minimal

HAND WATERING

Line sprinkler systems. Sprinkler systems are relatively inexpensive compared to other methods and can be used for irrigating both crops and pasture. Different types of sprinkler systems include hand line, wheel line, K-line, and solid set. Hand line systems usually require that pipe be moved, which adds to labor costs. It is possible to design a hand line system where all lateral lines are in place for the entire season, but this kind of setup takes careful planning and may reduce flexibility in cropping. The benefits of sprinkler systems include cost effectiveness and frost protection for crops. The component parts and setup costs for different types of sprinkler irrigation systems are shown in the table on the following page.

Irrigation guns. Irrigation guns are a specific type of sprinkler irrigation, most often used for irrigating large areas. Water placement is not precise and wind can alter the trajectory of water from the sprinkler, leading to uneven distribution of water across the field. Guns are best used for hay, forages, and large-scale plantings of a crop. (Note: If used on bare soil, the large droplets from irrigation guns can damage soil structure at the surface and cause compaction.)

Components: For irrigation gun, you need lateral pipe, gun stand or cart, gun, valve openers; for hard hose traveler, you need hard hose machine, gun, fill hose, valve opener

Setup costs: For irrigation gun, $280–$330 per acre; for hard hose traveler, $1,250–$1,500 per acre

Drip irrigation. Drip irrigation systems provide a very precise and uniform application of water at much lower pressure requirements than other systems. Water is targeted directly to the root zone of the crop, which improves overall efficiency. A drip tube runs the length of the row, and water is delivered through emitters in the drip line or through smaller tubes off the drip line that connect to mini- or microsprinklers. Drip systems are expensive and more complex than other types of irrigation methods. Advantages are efficient water use, reduced weed growth, lower energy costs, low labor requirements, and minimal maintenance costs. Drip systems also adapt well to fertigation. Fertigation, however, can increase the frequency of clogging, so drip systems used for fertigation should be set up to allow for regular flushing.

Components: Filter station, sub-main lines, control valves, controller, drip tubing, emitters, flush caps

Setup costs: For low-water-use crops, $1,100–$1,500 per acre; for high-water-use crops, $2,200–$2,700 per acre

Furrow irrigation. In furrow irrigation systems, a specially designed implement creates parallel channels that run the length of the field. Crops are planted on the ridge or narrow raised bed that runs between furrows. Gated pipe or siphon tubes deliver water from an irrigation canal into the furrows at one end of the field. The field is prepared to allow water to move evenly through it to promote uniform infiltration.

Components: Gate controls, gated pipe, siphon tubes

Setup costs: Varies by farm conditions

COMPONENT PARTS AND SETUP COSTS FOR FOUR SPRINKLER IRRIGATION SYSTEMS

TYPE	COMPONENTS	COST
Hand line	Aluminum lateral pipe, sprinklers, risers, valve openers, end plugs	$160–$190 per acre
K-line	Poly tubing, pods, sprinklers, camlocks, pull caps	$380–$450 per acre
Wheel line	Aluminum torque tube, wheels, sprinklers, levelers, drains, torque clamps, mover, fill hose	$550–$660 per acre
Solid set	Aluminum lateral pipe, sprinklers, risers, valve openers, end plugs	$1,120–$1,300 per acre

Water and Livestock

Livestock watering systems can be divided into three basic types: direct access, remote systems, and electric powered. The best type of system for a particular producer depends on the pasture layout, the amount of water required based on the number and type of livestock, the available electricity, and the source of the water. No matter which system you use, it is important to know the quantity of water consumed daily by the type of livestock you own and to account for fluctuations in water consumption due to daily and seasonal changes in weather conditions.

Direct access. Allowing livestock direct access to surface water can lead to a number of problems, including:

- Water contamination from stream- or pond-bank erosion and buildup of manure near the water source
- Poor herd health from exposure to water-transmitted diseases or animal injuries from slipping
- Uneven pasture utilization due to overgrazing near the water source and underutilization elsewhere

If direct access is the only option available, it is best to utilize controlled access points that allow livestock to drink without getting into the water source. This approach often includes hard surface ramps and fencing to protect the bank and provide stable footing for livestock. If considering this approach, it is best to consult with an expert (such as your local Cooperative Extension Service, soil and water conservation district staff, or private consultants) who can help you design the optimal system for your situation.

Remote systems. There are three basic types of remote systems: water hauling, gravity fed, and animal triggered.

Water hauling is used in intensive grazing systems, where livestock are sometimes moved daily to new pasture. Access to water can limit these moves. By utilizing a truck with a main storage tank and a portable stock tank, water can be continuously relocated throughout the pasture with the livestock.

In gravity-fed systems, small reservoirs or catchments can be plumbed to supply water for a livestock watering system, just like gravity-fed systems for crop production. For livestock, gravity systems generally consist of an uphill source (such as a pond) that flows into stock tanks equipped with float valves that regulate the delivery of water.

Animal-operated pasture pumps are operated by the animal pushing the pump diaphragm with its nose. These pumps can lift water a maximum of 20 vertical feet and, with the use of a pipeline, can also be set a quarter of a mile or more from the water source. These pumps are ideal for a livestock watering pond or for keeping livestock out of riparian areas. These pumps work well with cattle, horses, and pigs and can be adapted for use by sheep and goats.

Electric pump systems. Electric-powered water pumps are suited for streams, ponds, or wells and therefore provide the most options for getting water to livestock. These systems use a variety of pumps (depending on the

source of water) and easily connect to pipeline systems that deliver water to one or more stock tanks. If you are limited by the availability of electric power near the water source, or if you are looking to reduce energy costs, you might consider alternative energy sources. Pumping systems powered by alternative energy require water-storage reservoirs or stock tanks that can hold and deliver water to livestock between pumping cycles. Storage is generally sized to hold several days' worth of water.

Solar-powered systems. These are reliable and generally low maintenance. An array of solar panels collects and converts sunshine into electrical energy that is either used immediately to operate the pump or is stored by rechargeable batteries. Solar-powered systems vary in size and capacity. They can easily be scaled to handle the volume and flow that is necessary on hot, sunny days when livestock water needs increase.

Wind-powered systems perform best in areas that have higher-than-average wind speeds. Place windmills on high ground, where they have good exposure to the wind, and locate them far away from trees. Wind-powered pumps can pump from ponds, other catchments, and wells.

FARM PLAN

INFRASTRUCTURE AND EQUIPMENT ASSESSMENT

You can assess the equipment and infrastructure needs for your own farm or ranch using the Infrastructure and Equipment Assessment worksheet (appendix 1-14). To fill out the form:

- Begin with the list of current assets that you worked on earlier in this chapter (page 54).
- Continue working across the table, filling in what you think you will need and the most economical way to acquire it.

Farmers Share Their Experience

IRRIGATION

DEPENDING ON YOUR ACCESS TO WATER, irrigation may involve some of your biggest and most important investments in the farm. To summarize the key points from this irrigation systems and infrastructure section:

- Plan your irrigation system ahead of time and install a main line if possible.
- Thinking about efficiency at the outset will save you time and money in the long run.
- Make sure you have the water you need, and use irrigation methods that conserve water.
- Maintain your irrigation system in good working order to provide long-term benefits and lower costs.
- Understand the basic concepts of evapotranspiration, soil texture, soil-moisture monitoring, and soil water-holding capacity in order to manage your system for optimal water delivery to your crops and livestock.

The following insights from several farmers explain how they developed their irrigation systems: their access to water, how that water is supplied to their farm, and the type of system they use to get water to their crops and livestock. Some of the factors that influenced their decisions include water rights versus water availability, whether or not a water district is involved, how to pressurize the system (gravity fed or pump), and the most cost-effective way to apply water.

IRRIGATION LOGISTICS

Chris Jagger and Melanie Kuegler, Blue Fox Farm (profile on page 168)

Chris: Water rights are very important in Oregon and other areas of the West. In a lot of parts of the country, like where I grew up in Missouri, nobody knows what water rights are because water just comes from the sky. You can't farm here without having water rights. So to anybody looking to buy a piece of land in an area where water rights play a part, the first thing you should do is check that out.

Melanie: In this part of Oregon particularly, water availability can be a problem. And what that means is that the creek through our home

IRRIGATION POND

property actually runs dry. We didn't give as much thought to that as we should have before we bought this property. So we looked into water availability a lot more heavily when we were moving onto our other property. Not only does that property have really good water rights, there is also great water availability on that property through the summer.

IRRIGATION FILTER WITH FLOW METER

Mimo Davis, Urban Buds: City Grown Flowers (profile on page 129)

You have to irrigate here. We're on city water. We're metered. We pay by the gallon, so things have to be on drip irrigation or you will never succeed.

DRIP IRRIGATION

Carla Green, Sweet Home Farms (profile on page 216)

We purchased a small property nearby. It's 19 acres and it's irrigated. It's the only irrigated land that we have. We purchased that because we wanted good finishing ground during the heat of the summer. Everything else is nonirrigated. We figure our nonirrigated grass is done about the fourth of July.

Randy Kiyokawa, Kiyokawa Family Orchards (profile on page 94)

I am fortunate to live in an irrigation district that uses glacier water from Mt. Hood. I don't need to use pumps. I'm so fortunate for that. We use a lot of microsprinklers. The problem I have is not whether I'm going to have water, because we have such a good system up here, but how much pressure I will have and how much pressure will cause nozzles and other parts of my irrigation system to erode.

Sarahlee Lawrence, Rainshadow Organics (profile on page 218)

We get around seven to nine inches of rainfall a year, which is very little. Most of our moisture comes from snow. A big part of what we do in this area is take that snowmelt and store it and use it as irrigation water through our irrigation district from April through October.

IRRIGATION SYSTEMS

Randy Kiyokawa, Kiyokawa Family Orchards (profile on page 94)

The microsprinklers were mainly put in for conserving water. Also, they allow a grower to water more frequently. With the old hand sprinklers, it might take three weeks or longer before I rotated through the whole orchard. With the microsprinklers, I can water the whole orchard in a week and then maybe only get the top four or five inches pretty wet. It'll leach down. So I can get the water to a tree when it needs it instead of when I'm able to get around to scheduling changing the water.

Beth Hoinacki, Goodfoot Farm

(profile on page 44)

When we were building the farm, we started out with aluminum ag pipe for our irrigation systems. We didn't have enough to cover everything — we had to pick it up and move it if we wanted to water the other part of the field.

And so we chose to put the capital investment into PVC in the ground with sprinklers coming upright. We first did that on the blueberry field. Then we pulled the aluminum ag pipe out of the blueberry field and put it into our first vegetable cropping field.

Last year, we moved that aluminum ag pipe out of that vegetable cropping field and installed in-ground pipe with permanent sprinklers. The aluminum ag pipe will move to our next field that we're going to start cultivating next year.

Carla Green, Sweet Home Farms

(profile on page 216)

I think a lot of people forget about water development. But if you're going to have 60 little paddocks, you have to get water to 60 little paddocks. So we ended up laying a one-inch poly pipeline along our cross fences with hydrants every about 100 feet so we can run hoses to troughs pretty much anywhere in our fields.

Microsprinkler at Kiyokawa Family Orchards

In-ground pipe and permanent risers and sprinklers at Goodfoot Farm

FARM LABOR

Using our computer analogy from the beginning of this chapter, equipment and infrastructure create the framework for a farm — the hardware. It takes people to run it and make it successful — the software. Consider the following questions:

- What is your crop or livestock system?
- What kinds of equipment and machinery do you use?
- Will you be doing all of the farmwork or do you have a partner in the business?
- Will you need to hire employees?
- What kinds of changes might you expect when you hire employees?
- Do you plan to develop an educational program or work with volunteers?

Almost every decision you make in the farm business has some bearing on farm labor. For example, certain types of equipment and machinery can alter the labor requirements on the farm. In most cases, mechanization reduces labor costs. But purchasing a particular piece of equipment could also change the skill set that is needed on the farm (either to operate or to maintain the new machinery). When you assess equipment needs and purchases, you should take labor needs into account.

Further, your decisions about what crops to grow or what livestock to raise have a direct impact on your labor requirements. Annual vegetable crop systems (especially with a mix of crops and a staggered planting schedule) require constant attention throughout the growing season. Depending on the type of system, workers are usually involved in some aspect of ground preparation, seeding and transplanting, irrigation, weeding, harvesting, and marketing on a daily basis. Perennial crops, on the other hand, tend to require short, concentrated periods of labor at particular times of year. Production methods also impact labor needs. For instance, rotational grazing using electric fence takes more person-hours to manage paddocks and move livestock than a continuous grazing system involving a single pasture and supplemental hay.

Labor Needs

In making decisions about farm labor, it is helpful to imagine a continuum of labor needs:

Solo operation » Partnership » Interns and seasonal employees » Hired labor

Think about the different types of farms along that continuum. Where is your farm along this continuum and where would you like to be? The answer depends on your personal situation and preferences (what kind of labor setup you are most comfortable with), your plan for scaling up your farm, and the kind of crop or livestock system you are managing. Here are five different farm labor scenarios.

Solo operation (no help or hired labor). A small farm that you can manage on your own is one way to keep costs and time commitments under control. A one-person operation will be limited in scale, so it is important to retain as much of the gross revenue as possible for reinvestment in the farm and personal income. This type of farm will benefit from some mechanization to save time and effort, but it's crucial to minimize capital costs and infrastructure by using what is most efficient.

Partnership with family or business partners contributing labor. Many farms can be planned so that a family or partners can handle the workload. This scenario has scale limitations, of course, but if one of your goals is to keep the farm business small, then this could be one way to do that. It will help you focus your efforts on production and marketing and not on managing people. This scenario takes into consideration your lifestyle needs. Also, you will want to think about the division of workload within the family or partners.

Interns and apprentices. Some farms offer internships and apprenticeships to meet some of their labor needs. In some states it is illegal to hire unpaid interns or apprentices, but there are some legal options available for those who want to go this route and have the motivation to create an educational component to their labor management. One option is to pay interns and apprentices at least minimum wage and treat them as if they were employees. It is then up to the farmer to provide the educational experience, room and board (if applicable), or any other type of opportunities for apprentices. A second way to include interns is to work with an educational institution to help design the internship program and allow for interns to receive educational credit for working on your farm. Third, farmers can work with an established internship program that provides the intern and host farmer with curriculum, evaluation, and assessment models to meet education and training goals through a tuition-based model.

Seasonal labor. Many farmers need assistance only during certain times of the year, such as during harvest or when trees need to be pruned. Hiring seasonal labor is an option if you would like to keep your family as the main labor source while hiring extra help here and there when needed. The availability of seasonal labor varies by region and may be challenging. Some farms overcome this obstacle by sharing labor among more than one farm, thus creating a more attractive position for a farmworker.

Hired farmworkers as employees. Obviously, if the work becomes more than you can handle with your current labor setup, you will want to consider hiring employees. A common scenario is that the farm starts out as a family-scale farm and grows into one that hires employees.

It is important to be intentional about this process. Do your research and gather information about hiring practices and managing employees. Managing employees is not easy. A challenge can be how to best manage people in a way that keeps them happy about their work and that meets the farm's needs. Also, payroll may become a significant expense that will affect overall profitability. Look closely at the scale of the farm and the specific labor needs before hiring anyone. Why do you need to hire people? Is it really necessary, or could you continue operating the farm with family labor if you cut back on some crops, markets, and so on?

Hiring and Managing Employees and Seasonal Workers

Finding good employees can be challenging. Start with a clear and concise job description and then write a job announcement based on that description. Know what you are looking for in terms of skills, background, experience, and personal qualities. To advertise and get the word out about vacancies on your farm, the following resources may be helpful:

- Word of mouth (ask other farmers and let them know you are hiring)
- Email discussion lists
- Craigslist
- Ecological Farming Association (see Resources)
- Sustainable Food Jobs (see Resources)
- WWOOF-USA (see Resources)
- ATTRA's List of Internships (see Resources)
- Newspapers
- Social media
- State employment department job listings
- Job listings with local colleges and universities

Maintaining a good website for your farm business can also promote and generate interest in job openings you may have.

LABOR LAWS

When you reach the point where you need to hire employees, it is important to understand all the laws and regulations required of employers in your state. Some typical federal or state requirements for employers to hire farm employees include the following, adapted from "Checklist for Hiring Employees" (see Source Citations):

- Verify eligibility of employees to work in the United States (Form I-9).
- Obtain an Employer Identification Number through the Internal Revenue Service.
- Understand tax requirements and reporting for hiring employees. Records include W-4 and W-2 forms.
- Carry workers' compensation insurance for employees. Contact your state's workers' compensation division or department for more information.
- Develop a workplace safety program and understand what OSHA regulations you are required to follow.
- Find out what notices you are required to post at your farm informing employees of their rights and employer responsibilities. For instance, in Oregon the Bureau of Labor Industries helps with this.
- Pay at least minimum wage. Employees must keep accurate time sheets and sign them for validity.
- Determine if your agricultural business is exempt from unemployment insurance; this may be the case if your payroll is less than a certain amount in gross wages per quarter. Inquire with your local employment development office for more information.
- File your taxes in accordance with state and federal regulations.

BECOMING A GOOD MANAGER

Finding and hiring good employees is only a small part of the picture. To retain workers and help them be productive, you have to become a good manager. There are entire books and courses on this topic, but much of the wisdom from those sources can be distilled to these 10 rules.

1. Tell employees what you expect of them and put it in writing. If necessary, take steps to fire workers who ignore your requirements and expectations.
2. Make the best of each person's abilities.
3. Delegate and let go where you can, but stay involved and check in with employees on a regular basis.
4. Point out ways to improve, but never stand in anyone's way.
5. Give credit when it is due; look for unusually good performance and point it out.
6. Tell people in advance about changes that will affect them, and explain why changes will be made.
7. Remember that employees are often balancing work, family, and personal interests.
8. Compensate as well as possible. Don't grumble about making less money than your employees. It's your challenge to fix this situation.
9. Follow the Golden Rule: Treat others as you would want to be treated yourself.
10. Create incentives for your employees if you wish, but remember that no one will ever care as much about your farm as you do!

Farmers Share Their Experience
FARM LABOR

IN ANY FARM BUSINESS, PEOPLE ARE KEY. As Chris Jagger of Blue Fox Farm states: "People are really what make the farm go." But being a good manager requires some special skills. In the following profiles, note the different challenges farmers face and what solutions work for them. Also, look for the different priorities of each farmer and how those priorities continue to influence her or his decisions about hiring and managing farm labor.

MIMO DAVIS AND MIRANDA DUSCHACK, URBAN BUDS: CITY GROWN FLOWERS (profile on page 129)

Mimo: I'm the farm manager, so I decide what seeds get planted and where they go. And then we have an employee who has been with us for five years, and he takes everything I say and runs with it.

Miranda: We also have a part-time seasonal employee, and a full-time seasonal employee. I do farmers' markets, building and grounds maintenance, and presentations. At this point, and after becoming a mother, I can only work the farm 25 hours a week. Our next hire will be a secretary. That person will free Mimo and me to work more in the dirt and with plants.

URBAN BUDS

BETH HOINACKI, GOODFOOT FARM (profile on page 44)

For our labor needs, I've been the primary laborer. My husband has been the quintessential weekend warrior. It's also kind of a balance between management and grunt labor. We have to hire help. This past year, I had a full-time employee from mid-June until early to mid-August. And again, it was all about this idea of a system, and where does an employee fit into that system, and how does that make sense.

In terms of labor for a farm our size, I think we're going to need a maximum of two full-time employees, some seasonal employees, plus myself, so that I don't feel like I'm working too hard and getting stressed out.

GOODFOOT FARM

RENARD TURNER, VANGUARD RANCH (profile on page 132)

We run a small family farm. My wife and I do 75 to 80 percent of the work here. I have a friend who comes and helps sometimes, but we're all age 65 and over, so we are working smarter, not harder. We're doing a little bit less each year than we did the previous year, simply because the manpower does not allow for us to continue to expand, but also because expansion becomes your enemy at some point. The more help you need, the more money you have to pay out, the more materials you need, and the more your profit margin disappears.

When we started out, I decided I would grow organic produce for the underserved population, largely in the Richmond, Virginia, area and some in Charlottesville. And that did not work out very well because it required a lot of labor.

The unfortunate situation for most of us African-American farmers is that it's very difficult to find laborers. We have a very hard time trying to find interns largely because there is anti-agricultural backlash due to the legacy of slavery. Many African Americans see agriculture as basically a backward movement and not a progressive, forward movement. But it's exactly the opposite, because it's way different when you're the landowner. It actually liberates you in ways that most people don't even realize.

SARAHLEE LAWRENCE, RAINSHADOW ORGANICS (profile on page 218)

I'm an only child. I live here with my parents, my grandma, my aunt and uncle, and my husband, Ashanti Samuels. We have one full-time employee, and we host many WWOOFers — Willing Workers On Organic Farms.

RANDY KIYOKAWA, KIYOKAWA FAMILY ORCHARDS (profile on page 94)

We have employees that come back year after year. I'm fortunate enough to have some staff who have been with me since I was in middle

Renard Turner handles the majority of farmwork at Vanguard Ranch.

WWOOFers (Willing Workers on Organic Farms) are a source of labor for Rainshadow Organics.

school. Some of the new guys have been with me for 10 or 20 years, so I don't have a lot of turnover — it makes my job a lot easier.

CHRIS JAGGER AND MELANIE KUEGLER, BLUE FOX FARM

(profile on page 168)

Chris: Nine months out of the year, there's between five and eight of us. And then the other three months out of the year, I keep a crew of just two and myself. At the height of the season, we work from 6:00 in the morning to 5:00 in the evening and rarely any later than that because we want to be home with our kids in the evening so they know their parents.

Melanie: I only physically work on the farm two days a week. Because we have two sons that are very young right now, I'm spending my other days with them. I am doing all the bookwork. I pay all the employees. I do all the bills, invoicing, and record keeping. Winter is probably when I'm the busiest with all of that, because I'm gearing up for the next year. I'm also taking care of all the propagation greenhouses. Since they're here on our home property, I can do all of that work.

Chris: We started out having just interns. It has become more and more challenging to legally deal with having interns, and to make sure people come away from the experience with a good taste in their mouth. It's hard for the interns to really see the full spectrum of what we do as a farm, because people are slotted into more specialized roles.

Melanie: Because of the difficulties around having interns, we've ended up hiring employees and we're really happy with that. When you have somebody for multiple seasons, you create a relationship with them. You better understand their work. They better understand the seasons so they're more efficient workers, and we're all a lot happier. So our goal is to pay them properly and make sure that they're happy so they stay with us.

Saul Heidia (left) and Randy Kiyokawa (right); Saul has worked at Kiyokawa Family Orchards for more than 40 years.

Blue Fox Farms has up to eight employees during the growing season and two during the winter.

LABOR PLAN

The Labor Plan worksheet (appendix 1-15) helps you evaluate your current situation and where you want to be in five years. As part of your labor plan, it may also be helpful to create a calendar of work for the farm to visualize what kind of work is needed and when.

Farm Labor, in Summary

You know your situation and the needs of your farm business better than anyone. Make informed decisions based on that knowledge and insight. Start small in your first year and grow gradually, with thought and planning. Hire employees if workload demands; do not count on doing it all yourself. But know that there are definite risks involved in hiring labor. Be aware of those and go forward with a realistic outlook. Here are some of the main issues:

1. **Labor availability.** It may be hard to find qualified and experienced farmworkers in your area or you may feel like you can't afford to pay workers.
2. **Work quality.** Work must be completed on time and done well.
3. **Nonwage expenses.** You could incur high indirect labor costs. This is usually associated with high turnover and absenteeism, which results in spending a great deal of time orienting, training, hiring, and terminating employees.
4. **Internal conflict.** This can manifest as quiet disgruntlement that takes its toll on employee performance, or in more serious instances can end in litigation.
5. **Time.** Employee supervision, management, and training require time.
6. **Legal infraction.** This is the risk of violating a law or regulation and having to spend time and money dealing with those problems.

FARM ENERGY

The foundation of all the operational aspects of your farm is energy. Consider how energy is expended, where it comes from, and how it flows. There is the solar energy on which plant and animal production is based (photosynthesis). There is the energy required to run equipment, till soil, move water, heat greenhouses, and apply nutrient and pest control inputs. You also need energy to harvest, store, process, and transport your farm's product. Add all this up, and you have a complex subsystem within the farm that comes with a high price tag.

Reducing the cost of energy expenditures and making the most of what you do spend on energy are major factors in building a successful farm business. How do you improve the energy equation on your farm? Is it possible to reach energy self-sufficiency in a farm business?

Energy Audit

An energy audit can be a helpful first step in addressing energy issues on your farm. Such an assessment should account for energy needs and energy consumption, look at ways to improve energy efficiency, and find alternative energy sources appropriate for your budget. Various state agencies and nonprofit organizations provide information on energy audits. Some states have organizations that conduct or assist with farm energy assessments. Contact your local power company or Cooperative Extension Service for resources in your state. If you want to conduct your own assessment, there are farm energy calculators available to help you with that task. Learn more with ATTRA's Farm Energy Calculators (see Resources). Once you have completed this assessment, you can pinpoint areas to improve energy efficiency and begin to look at alternatives to reduce overall energy costs on the farm.

FARM ENERGY ASSESSMENT

To start you thinking about this important facet of your business, work on the Simplified Farm Energy Assessment worksheet (appendix 1-16). This worksheet will help you with a big-picture analysis of your farm's energy needs and consumption.

Alternative Energy

Petroleum-based energy sources have climate and environmental consequences and may become too costly in the future. Developing alternative energy sources could help improve the energy equation on your farm. There are many grant and loan programs promoting alternative energy that can help you pay for installation of (or retrofit) a sustainable energy project on your farm. Cooperative Extension Service and NRCS offices in your state also provide information for farmers on energy alternatives. Four alternative energy options are summarized below.

Solar. Photovoltaic systems (solar panels) can provide energy for a variety of farming needs: pumping water to crops and livestock, refrigeration, air cooling and heating, water heating, electric fencing, lighting, and greenhouse heating. Solar panel systems are the most common way to produce on-farm renewable energy in agricultural systems. Financial support is available from various sources to install systems on farms. An important consideration in solar systems is assessing your solar potential and understanding solar gain throughout the year. Passive solar design (a lower-cost option than photovoltaic systems) can also be used for greenhouses and other farm buildings. To learn more about passive solar greenhouse design, watch the video *Thermal Banking Greenhouses* from the USDA SARE program (see Resources).

Hydro. Microhydropower systems use the energy of flowing water to produce electricity. A run-of-the-river system is the most common type used for microhydro systems. In this type of system, water is diverted to a channel or pipeline that delivers it to a waterwheel or turbine. Microhydro systems are low impact and nonconsumptive and can last 20 years or more. This may be a viable option for those who have access to water sources that flow 24 hours a day. To learn more, watch *Microhydropower Systems* from the U.S. Department of Energy (see Resources).

Farmers Share Their Experience

CREATING EFFICIENCY

AS YOU READ THESE INSIGHTS from farmers, note how each farmer manages the interface between equipment, people, and energy. The common theme running through each story is creating efficiency, and an important aspect of this theme is that efficiency doesn't happen automatically; it has to be created by you, the farm manager.

GOODFOOT FARM

BETH HOINACKI, GOODFOOT FARM (profile on page 44)

I can be out there working in the field and thinking, "This doesn't make sense to me. There's got to be a better way. This is ridiculous." And there usually is a better way. It comes down to efficiency. If Adam is tilling and he must make a five-point turn to do the next row, that extra time adds up over the number of years we're farming, and then you see that having too small a headland space is costing us money. So that's why we ask ourselves, "How can we be doing this better?"

Before we remodeled a barn to be used for pack-out, I used it one summer for getting ready for market. I was able to figure out how to move most efficiently in the space, so that when we ramped up production, everything was in the right spot. We had the walk-in cooler exactly where it needed to be, we had the sinks right where they needed to be, and the washtubs were in the perfect location.

SLOW HAND FARM

BLUE FOX FARM

SWEET HOME FARMS

JOSH VOLK, SLOW HAND FARM

(profile on page 46)

I feel a little silly sometimes because I have a lot of different tools that do very similar tasks, but sometimes one tool is more efficient than another, and this is especially obvious when working by hand. For instance, some tools work better when it's wetter, and some tools work better when it's drier and the soil is a little bit harder. Depending on the soil conditions and what the crop is, there might be one tool that's cultivating really well early in the season and another tool that's cultivating better later. For this reason, I have a variety of different hoes that I use for cultivation, as well as a variety of forks and a spade and a couple of different rakes that I use for bed preparation.

CHRIS JAGGER, BLUE FOX FARM

(profile on page 168)

We definitely go by the philosophy "work smarter, not harder." Just this last year, we got a transplant sled, which has really made us more efficient. We used to spend a full day transplanting; now we usually finish in three or four hours, and the planting is more accurate to boot.

We've also bought enough handset sprinkler pipe that we don't have to move pipe once we've set it. That's paid for itself in no time.

CARLA GREEN, SWEET HOME FARMS

(profile on page 216)

Before we put in permanent fencing, we used polywire temporary fence to set up lanes and cross fences to test things out. We did end up installing some interior cross fences as well as a lane, and we can easily put in polywire to create small paddocks so that we could manage our grass well.

Farmers Share Their Experience

HORSE-POWERED FARM INFRASTRUCTURE: GOOD WORK FARM

A UNIQUE ASPECT of Good Work Farm (profile on page 92) is that most of the work requiring power is accomplished with horses. This approach is consistent with the farming traditions in the region and fits closely with Lisa and Anton's values.

Anton: We're in a unique position in that we're using horses, so most of the equipment that we have is for them. We're lucky that we're not far from Lancaster County, where there's a large Amish and Mennonite community, so there is a large amount of used and newly manufactured horse equipment in our area. For the most part, we have gotten used or rebuilt equipment. Most of our farm is in single rows on 36-inch centers that the horses can cultivate, but we do have a Kubota tractor for general farm use, like turning compost and mowing.

We were the most nervous about the plowing equipment because it takes a good degree of skill to properly adjust a plow — it's a real art. As soon as we got the horses, we purchased a new riding plow so that we knew all the parts were there and all the pieces were working properly. We employ a used disc and a used spring tooth harrow for most of our secondary tillage. And then we felt that we were falling behind when we did the primary tillage with the horses, so we purchased a used chisel plow that we use with a borrowed tractor.

We use a rebuilt sickle bar mower with the horses for cover crop management and farm field mowing. We also got a brush hog to maintain field edges if we get behind on mowing with the horses.

Working with the horses is very satisfying. One of the main reasons I got into farming was to try to reduce my impact on the environment. Horse farming really slows down the pace of life and causes you to pay more careful attention to what's happening on your farm. I really like that element of draft-powered farming.

Lisa: The more I learned about farming and how to farm, the more I wanted to farm with horses. They made sense together for me. And I love working with horses.

It also doesn't come without struggle. Since almost all of our equipment is set up for horses, we are in a hard situation if we can't use them for any period of time. In the past, we've gone through weeklong stretches where we couldn't work horses due to injury and had to make do with what we could. And in terms of the quality of life, we have to find somebody to take care of the horses if we are going away.

We are continually reassessing whether we want to be farming with horses. But I think the answer is always yes, both because we love it so much and also because we feel so committed to the ethical choice of growing food with less fossil fuels.

Wind. In locations where there are enough windy days to justify the expense, small wind electric systems are extremely effective for the small farm. These systems harness the kinetic energy in the wind through turbines and generators. They can be used for all electricity needs as well as pumping water. In rural communities, cooperatively owned wind projects are also an option. It's important to assess the site's wind energy potential before installing a wind electric system. It may be possible to link with a larger wind installation firm that will pay to install the technology on your farm.

Biofuels and bioenergy. Ethanol and biodiesel are two alternative plant-based fuels that farmers sometimes use for alternative energy. When space and scale allow, farmers are growing the fuel for their own farms. Biomass alternatives are also becoming increasingly popular ways to generate energy. Wood by-products are burned, producing steam that turns a turbine to generate electricity. Farmers can access these types of energy in various ways. To learn more about biodiesel production, watch the video *Energy Independence: On-Farm Biodiesel Fuel Production* (see Resources).

On-Farm Energy Efficiency

Whether or not you decide to pursue a full-fledged alternative energy project for your farm, there are a number of ways you can reduce your on-farm energy use. Look for ways to apply energy-efficient technologies, such as:

- Energy-efficient lighting
- Solar hot water heaters
- High-efficiency motors and fans
- Best management practices for fuel storage, equipment maintenance, and field operations to reduce fuel consumption

Making changes in your production system can have a major impact on energy use. For instance, raising livestock in a grass-based system, using draft horses, or installing drip irrigation can all reduce energy consumption, which reduces costs and increases the financial viability of a small farm. The long-term sustainability of the farm also improves as pollution is reduced and there is less reliance on fossil fuels. For more information on farm energy efficiency, visit the USDA NRCS Farm Energy Efficiency website (see Resources).

PUTTING IT ALL TOGETHER

In this chapter, you have assessed the infrastructure and equipment needs of your farm, looking particularly at different options for season-extension techniques and setting up an irrigation system. You have thought about the people side of the business and learned about the issues and decisions involved in finding, hiring, and managing employees. And you have evaluated your farm's energy usage and looked at opportunities for reducing your carbon footprint.

Take a moment to reflect on what you have learned in this chapter and think about how it connects with other aspects of your farm. Look for ways to integrate the concepts, information, and activities in this chapter into your whole farm plan, and as you read the next chapter, think about ways that information can be included in your decisions about marketing and production practices.

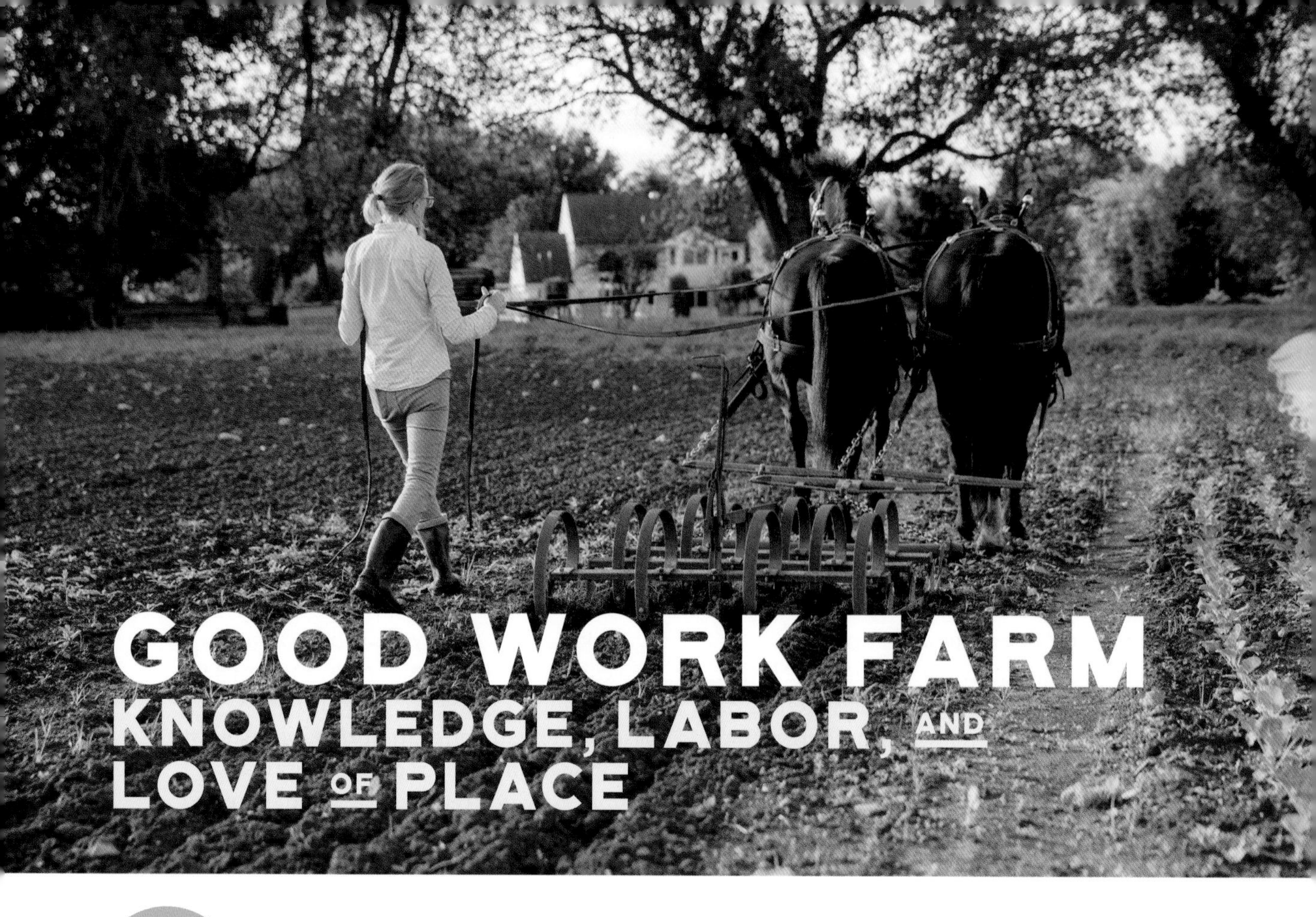

GOOD WORK FARM
KNOWLEDGE, LABOR, AND LOVE OF PLACE

Good Work Farm is a 14-acre diversified farm located in Pennsylvania's Lehigh Valley. Owned and managed by Anton Shannon and Lisa Miskelly, the farm produces vegetables, fruits, herbs, and flowers for its expanding CSA.

Anton Shannon, a graduate of Pennsylvania State University's Beginning Farmer Training Program, started the farm back in 2011 with colleague Sarah Edmonds, renting land from an incubator farm program called the Seed Farm. After Sarah moved on from the business after a couple of years, Anton met and went into partnership with Lisa Miskelly. Anton and Lisa rented land from a private landowner for several years, always with the hope of eventually purchasing their own land. After a long search, they found land that would work for them and the kind of business they wanted to develop. "After operating the farm on rented land for a number of years, we are thrilled to be gearing up for the first season on our own property," says Anton.

Anton and Lisa come to the farm business from different backgrounds: Anton grew up in the Lehigh Valley and was drawn to farming in his study of the social, economic, and environmental implications and interactions of land use; Lisa, born and raised in suburban Maryland, had experience working on a variety of small farms in Vermont and other parts of the Northeast. Their partnership is based on a shared vision for the farm and their heartfelt commitment to the land, conservation, farming, and their community. "We work to create a strong, honest, transparent, and

mutually beneficial relationship between farmers and consumers," notes Anton.

Good Work Farm has been growing crops according to organic standards for the past six years, but it is not yet certified organic. "We purchase organic seeds, and we use only organically approved fertilizers, amendments, and materials," says Anton. "We also use crop rotations and practices that reduce erosion, build soil fertility, and improve soil health." With the purchase of their own land, Anton and Lisa now plan to start the certification process.

Anton and Lisa currently market most of their produce through a 55-member CSA and are in the process of developing a farm stand at their property: "We have a really strong CSA community. Our customers support our work because they value the role farms play in their community and care about the taste and nutritional content of their food. Share members have the opportunity to regularly walk the land, experience the rhythms of farm life, pick their own peas and strawberries, and watch their farm progress and thrive. They don't just pick up a box and leave."

One of the farm's most unique aspects is that it is mostly horse powered, a passion shared by both Anton and Lisa. In fact, they met at a horse farming workshop held in Cortland, New York. "Horses contribute to our farm by supporting healthy soil life, providing fertility, and decreasing our dependence on foreign sources of energy," notes Anton. They currently have two draft horses — Daisy and Duke — used for primary and secondary tillage. Anton says, "We use the horses as much as possible and use the tractor whenever necessary." They maintain the horses on a couple acres of pasture that they are developing on their own property, as well as some rented pastureland. Says Lisa, "I feel like every step that we take toward farming with horses is another step further on our path. At one point, I think our end goal was to be farming 100 percent with horses. That may not be realistic, but it's still something that we hold dear, and greatly esteem those farmers who are choosing to farm that way. I don't know that it will ever fully fit our life. But it's certainly a strong value for us."

Good Work Farm takes its name from an essay by Wendell Berry entitled "Conservation Is Good Work." In that piece, Berry writes how good farming depends on traditional knowledge, physical labor, and love for one's place on earth. In the spirit of that essay, Anton and Lisa note that what matters most is the farmers' commitment to their work, their land, and their community. "Farming provides us with a livelihood which fulfills our physical need for healthy food that retains its connection to its source," remarks Anton. "At Good Work Farm, we try to farm well through hard work, careful planning, paying attention, and constantly learning. We farm to awaken our connection with the earth, the elements, and the unknown."

KIYOKAWA FAMILY ORCHARDS

FARMING AS A FAMILY TRADITION

Randy Kiyokawa's family has been farming in the Hood River Valley for over one hundred years. His grandfather came from Japan in 1906 and started farming in the area in 1911.

In 1951, Randy's dad came back to the Hood River Valley and purchased the current farm through the G.I. Bill. Randy grew up on the farm, but it wasn't until 1987 that he also returned to farm full-time and manage the business. Randy describes how he grew into the farm manager role: "Starting out, I was a laborer and did everything," he says. "But I didn't make any of the decisions. Over the next several years, I was allowed to make more and more decisions, for better or for worse."

Kiyokawa Family Orchards has about two hundred acres on which Randy grows apples, cider apples, pears, peaches, cherries, plums, pluots, and pluerries. The main crop is pears, which are sold through Diamond Fruit Growers. But the farm also sells directly to consumers at a farm stand and at 15 separate farmers' markets in the region. "The farm is different from when I grew up back in the '60s and '70s, in that we do a lot more direct marketing instead of just wholesale," says Randy. "One of my greatest joys is the

Top middle: Randy Kiyokawa (3rd generation) with Korbi (5th generation); **Top right:** Michiko Kiyokawa (2nd generation); **Bottom left:** Catherine Kiyokawa (4th generation)

customers I have direct contact with — the customers that say, 'Boy, this is the best apple I've ever eaten.' That and the employees that come back year after year."

Randy's grown children now have the opportunity to return to the farm with their own families. "What drives me is fear, fear of losing the farm. I don't want to be the generation that loses it. Literally. It's hard. The responsibility I have, not only to my family but to my employees, is a concern for me. That's what makes me wake up every morning and make sure I do what needs to be done."

"What drives me is fear, **fear of losing the farm. I don't want to be the generation that loses it."**

— RANDY KIYOKAWA

SELL IT

RAINSHADOW ORGANICS

CHAPTER THREE

MARKETS AND MARKETING

This chapter will help you think about your marketing options and understand how marketing decisions are interrelated with other aspects of the farm business. Your mission statement, vision, and values, as well as the other assessments you worked on in chapter 1, "Dream It," will be important in developing your marketing strategy. As you will discover, marketing decisions are linked closely with which crops you grow, what livestock you raise, and your production practices. Many marketing decisions need to be made well before the start of the growing season. Marketing decisions are also related to budgeting and cash flow needs throughout the year, so look for connections with the management topics covered in chapter 4, "Manage It."

We begin with some basic marketing concepts and then address some of the most common marketing channels small farmers in the United States use. The information covered here will assist you in deciding which marketing options work best for you.

BY THE END OF THIS CHAPTER, YOU WILL:

- Know the key concepts and philosophies that form the foundation for successful farm marketing.
- Understand the different types of markets — and the challenges, benefits, and considerations associated with each.
- Identify which marketing strategies are most compatible with your farm business.
- Know what information you need to create a successful marketing plan.
- Connect the mission of your farm to your marketing strategy and financial goals.

WHAT IS A MARKET?

A *market* is any group of potential customers that forms for the purpose of selling and purchasing a product. A *marketplace* is the space in which products and services are offered and purchased. It can be a farmers' market, the Chicago Board of Trade, or the Internet. A market is sometimes synonymous with *marketplace*. Markets can be classified by area (local, national, or international), by type of marketing channel (wholesale, retail, or direct), or by the type of customer (individuals or other businesses). *Marketing* is the action a farmer takes to engage customers in the marketplace. How would you define the marketplace for your farm business?

Types of Markets

To simplify our discussion, we divide markets into two broad categories: wholesale markets and direct markets.

WHOLESALE MARKETS

In this category, you sell a product through a wholesale market or distribution channel. Wholesale markets include a variety of brokers and agents who buy, sell, and handle products. Examples include local livestock auction yards, community grain elevators, grower co-ops, canneries, and wholesale distributors. Wholesale markets are used for both fresh and processed commodities, most often deal with large volumes of products, and include large storage capacity and extensive transportation systems. Sales to restaurants, local retailers, and institutions are often technically included within definitions of wholesale markets, but for our purposes we consider them direct markets because farmers have an influence on the price they receive. We'll cover more on this in the Direct Markets section later in this chapter.

You have little say in what you can charge for your product in a wholesale market situation

Examples of direct markets, clockwise from top left: Vanguard Ranch concession stand, Kiyokawa Family Orchards farm store, Rainshadow Organics farmers' market display, Rainshadow Organics farm store

since the price is determined largely by the chain of players above you in the hierarchy (farmer to wholesaler to processor to retailer), who are responding to supply and demand in the marketplace. Although more than 95 percent of farm products pass through wholesale market channels, many farmers choose not to go this direction. Wholesale marketing has both advantages and disadvantages.

ADVANTAGES OF WHOLESALE MARKETING

- Wholesale marketing requires minimal time and effort. Your main responsibility is to deliver the product to the wholesaler in a timely manner. You may then focus on farming, while someone else in the chain takes care of marketing.
- Agreements with wholesalers can provide you with a predictable and consistent market for your product, giving a sense of stability.

CHALLENGES OF WHOLESALE MARKETING

- Wholesale markets typically offer a low per-unit price, so large volumes of product must be produced to generate any potential profit.
- Small-scale farms may have insufficient volume to participate profitably in wholesale markets.
- You have little say in what price you get for your product.
- Wholesalers may require you to have third-party certifications addressing production practices (organic) and food-safety plans (Good Agricultural Practices [GAP]), as well as liability insurance.

As a small farmer, you may be wondering if you should ever consider wholesale markets. Certain wholesale markets can be appropriate and useful for small farms in particular situations, especially for those who are scaling up their operation or expanding their product line. As production increases, you may find that you cannot sell all your product through your usual direct market channels, in which case wholesale markets provide another outlet. Wholesale markets also present another source of cash flow, which can be important during certain periods of the year. The two critical questions that you need to ask are: Do I have the type of product the wholesaler is looking for? Can I provide that product in the quantities required by the wholesaler?

MARKETING CHANNEL SELF-ASSESSMENT

An important step in developing your marketing strategy is to think about your personality traits and lifestyle goals — and how they relate to the type of market channel that will work best for your situation. Work through the Influence of Personality and Lifestyle Goals on Marketing worksheet (appendix 1-17) to begin that process.

DIRECT MARKETS

Direct marketing is when you sell a product directly to customers, such as at a roadside stand or farmers' market. You are more involved in the direct marketing process than in wholesale marketing, and it requires a different set of skills. You are not just growing a crop, you are producing a product that must be presented to customers in a way that makes the customers want to buy it. The purchase experience itself is often just as important to the customer as the product. Over the past several decades, as consumers have become more interested in purchasing directly from farmers, direct marketing has been growing.

The direct market channels we explore in this chapter include farmers' markets, community-supported agriculture (CSA), farm stands, U-pick, and agritourism. We also include sales to restaurants, retail stores, and institutions because the farmer has some control over the price and the transaction is based on a direct relationship and interaction with a business owner or manager.

The price you receive varies by specific channel and locality. Often, you receive a price similar to retail. For sales to restaurants, grocery stores, and institutions such as schools, prices may be below retail and are often closer to wholesale, with prices within these channels higher for sales to restaurants and lower for sales to grocery stores and schools. Direct marketing also has a number of advantages and disadvantages, as shown below.

As seen in the illustration opposite, direct markets may target the customer directly or indirectly through a retailer such as a grocery store or restaurant. Wholesale market channels may be simple with few intermediaries (wholesaler to retailer) or more complex with multiple intermediaries (addition of brokers and other intermediaries).

The takeaway message here is be strategic in your choice of marketing channels. Finding the right mix of wholesale and direct marketing channels may be the best approach and may spread risk as you work through the transition.

ADVANTAGES OF DIRECT MARKETING

Direct marketing is often the best option for farmers who deal in small quantities of product.

You set the price (or at least have more control over the price), so production costs and profit margin can be taken into account.

Payment is often immediate, and good products and services can get attractive prices.

You receive instant feedback from customers on products and services. Based on this feedback, you can improve your business practices and potentially increase farm profitability.

CHALLENGES OF DIRECT MARKETING

With direct marketing, you are responsible for what is grown, and how and to whom the product is marketed, so risk is generally higher than it is for wholesale markets.

Direct marketing amplifies the small-business aspect of your farm: you take on new roles and become responsible for marketing, retailing, advertising, customer relations, and so on.

Direct marketing requires people skills and a high level of motivation to develop and maintain customer relationships.

There are regulations that pertain to direct marketing that farms selling into wholesale markets do not worry about (such as licenses, packaging, weights, and measures; see chapter 6, "Keep It," for a discussion of these topics).

Many direct markets have limited capacity. For instance, is there adequate demand for an additional vendor with your products at your farmers' market?

Even though the potential for profit is much greater, direct marketing requires many additional hours (beyond growing the crop or raising animals) to market and sell your product.

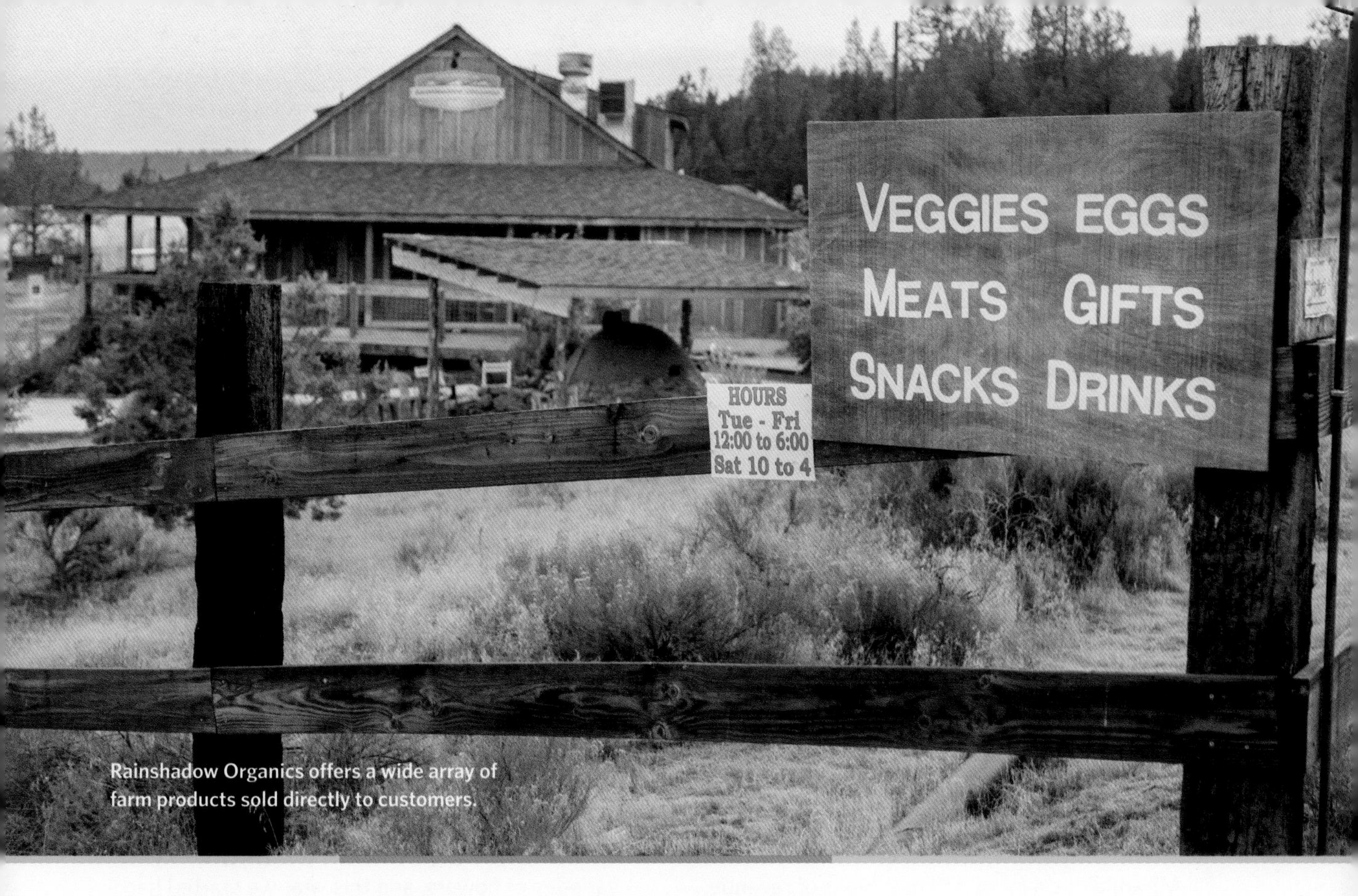

Rainshadow Organics offers a wide array of farm products sold directly to customers.

4 WAYS TO MARKET YOUR PRODUCT

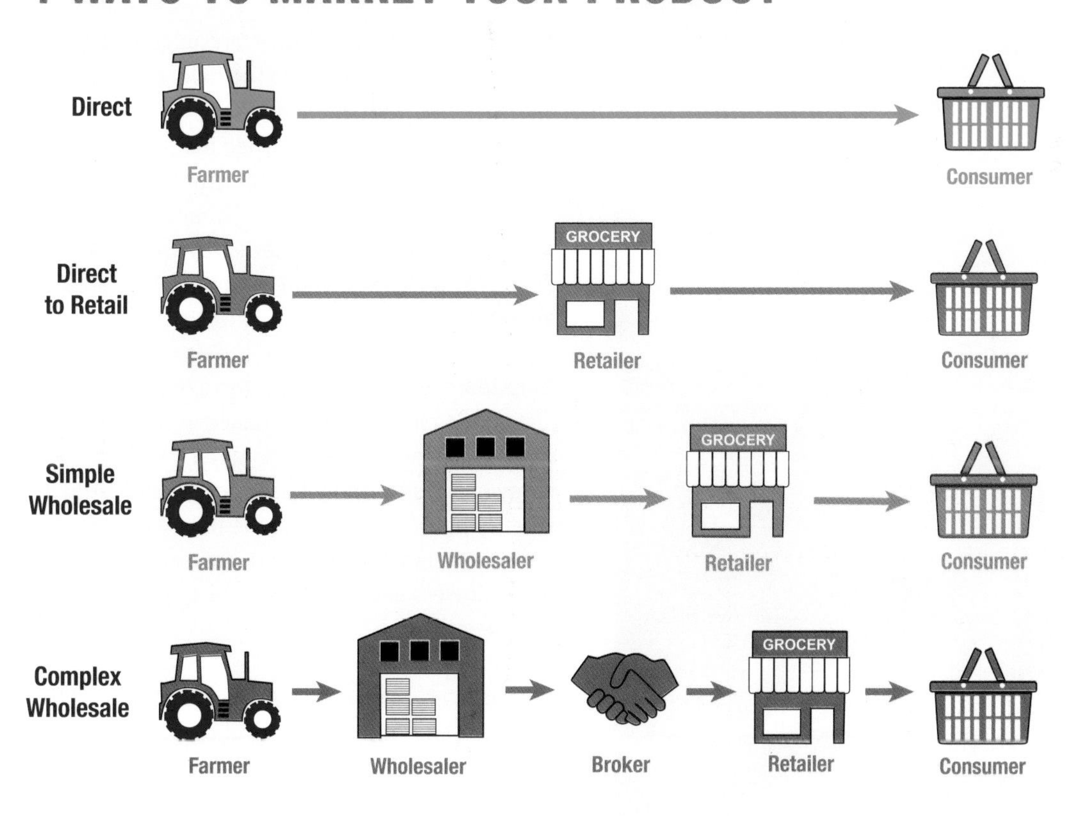

Farmers Share Their Experience
ENVISIONING YOUR MARKETPLACE

WE ASKED HOW YOU MIGHT DEFINE the marketplace for your farm at the beginning of this chapter. Note how the farmers from Urban Buds: City Grown Flowers in St. Louis define the marketplace for their farm: their customers, the geography and locale in which they work, and the connections with people and the wider community.

MIMO DAVIS AND MIRANDA DUSCHACK, URBAN BUDS: CITY GROWN FLOWERS (profile on page 129)

Mimo: When we first started out, I thought, if I'm going to be an urban farmer on this limited lot, I'm going to put on that lot whatever is going to give me the highest dollar for the square footage. And that's flowers. We sell direct to florists, we do two farmers' markets, and we do some weddings, events, and tours. We're also developing an agri-education curriculum.

Miranda: It's a lot of different types of revenue, but we're dynamic women, so our business reflects that.

Mimo: We're in the beautification business, and we're in the business of honoring life's really important things. Flowers are with you when you're born, when you're married, and when you die. That's so important.

Miranda: People view flowers as a luxury good, but we need roses just as we need bread. People are always going to have flowers. We grow 70 varieties of flowers, and this year we started harvesting on January 1.

Mimo: So when most flower farmers are just gearing up and getting into the field, we're almost halfway done with our season.

The floral holidays are when it's cold outside. If you're not in the marketplace for those holidays, you're losing a chunk of change. Where we are, you can't hit Mother's Day with just field-grown flowers. There's no way. You need a greenhouse.

Miranda: If you don't do season extension, you're just coming into the marketplace in June, when many customers leave on summer vacation and many university students are out of town. And it is really hot. August is also the time when vegetable farmers who decided to have an extra row of sunflowers and zinnias are starting to move into the market.

Mimo: So my whole plan is to flip that and grow on what I call "the other side of the calendar." I

still think the farmers' market model is good. A lot of people don't seem to think that, but we love it. We do very well at our farmers' market. On Instagram, I posted a thank-you note to our customers this past week. They sent me Instagram posts of their flowers, and that made me realize that I go home with these people every Saturday.

Miranda: A little piece of us goes home with these people, and then we live with them all week. It's no wonder we're so tired.

Mimo: That relationship is amazing. Those flowers sit on their table all week long. They look at them and they think about the farmer, and then those people come back every Saturday. How special is that? I don't think there are a lot of farmers out there who really get how important that relationship is. Because if you get how important that relationship is, you wouldn't be shy about your prices. You wouldn't be shy about your product. You would have respect for your product and your market and charge its value.

Yes, I love the flowers. Yes, I love people's reactions when they see our 70 amazing, beautiful varieties of cut flowers. But none of that pays the bills, right?

So I demand a high price.

Miranda: There's more and more competition now, but also more interest.

Mimo: We had a lot of feelings about competition, but this year we've moved into embracing it. If people are going to want to learn about cut flowers and pay gobs of money, they can pay me.

Miranda: And so that's why we've opened up to on-farm workshops. We let this competition fear go, because they're coming anyway. I'm very proud of Mimo and how she is a pioneer in this.

We want to continue taking on weddings, but we're reevaluating how much energy we want to put into that aspect of the business. If we could do it well, it would be the most lucrative revenue stream, but it is very hard to do well. Anyone who's reading this book and wants to become a farmer florist, I would caution you to have a very detailed business plan. Get some specific business training in being a florist.

Mimo: They're really two separate businesses.

MARKETING STRATEGY

A basic marketing strategy addresses three key questions: how, what, and when to market. To answer these basic questions:

- Explore the marketing mix framework (how to market).
- Identify enterprises and products (what to market).
- Consider the best timing — and time frame — for your products (when to market).

How to Market

Marketing is a major component of developing and managing your business. At the beginning of this chapter, we defined marketing as the actions you take to engage customers in the marketplace. Those actions can be summarized as the marketing mix — the tools you use and choices you make to bring your product to market. The marketing mix for a particular business is often described by the Four Ps: product, price, promotion, and place (adapted from *Basic Marketing*; see Source Citations).

Product. A product is anything that satisfies a customer's need or want. It includes elements such as the physical appearance of the product, packaging, branding, and quality guarantees.

Price. Price conveys the usefulness of the product in money terms. Often determined in the marketplace, pricing is important to both you and the customer. For you, cost of production and profit margin are key considerations in determining the target price. For the customer, other factors such as quality, freshness, similarly priced alternatives, and access determine the price they are willing to pay.

Promotion. Promotion is communication with customers to motivate them to purchase the product. It includes advertising, sales methods, product display, signage, and branding.

Place. Place or placement describes how the product will be distributed and available to customers. Place also refers to the marketing channel (direct, retail, wholesale), the geographic scope of the market (local, regional, national), and access options (in person, online sales).

The Four Ps have been used in business education since the 1960s. More recent thinking has brought a stronger customer focus to the marketing mix. This newer approach requires you to know your customers' needs, sense marketplace changes quickly, and be nimble in responding to changes. The current approach is summarized by SIVA (adapted from "In the Mix"; see Source Citations):

- **Solution:** A customer need is solved.
- **Information:** A customer must have information in order to make a purchase.
- **Value:** What will be the cost, and does it equal the benefits?
- **Access:** Where can the customer easily make a purchase?

When aligned with the traditional Four Ps, the customer perspective adds greater clarity to your marketing strategy:

- A **product** is a **solution** to a customer need.
- **Promotion** provides **information** to make the customer aware of the product.
- **Price** is the **value** of the product to the customer.
- **Place** is how and where the customer may **access** the product.

The chart below shows how the Four Ps and SIVA might play out in two different marketing examples. For each example, note the many decisions involved in that aspect of marketing the product.

PRICING

Establishing the price for your product is a key step in reaching your financial and business goals. The challenge is in finding the sweet spot — the middle ground between a price that is so low that you are not covering your costs and one that is so high that sales drop off. Farmers usually follow one of the following two strategies for setting prices.

Competition-based pricing, where the price is set largely based on market conditions. This strategy requires a knowledge of your competitors' prices and understanding what the market will bear.

Cost-based pricing, which factors in the true production costs including overhead costs. In this pricing scheme, you set prices to cover your costs, including owner's draw and profit.

TWO EXAMPLES OF SIVA AND THE FOUR Ps

SOLUTION (PRODUCT)	INFORMATION (PROMOTION)	VALUE (PRICE)	ACCESS (PLACE)
Heirloom, vine-ripened tomato	Attractive and prominent signage at farmers' market	A price per pound based on production cost plus a quantity discount for 5 or more pounds	3 farmers' markets per week
Grass-fed beef	A farm website; advertising in a regional newspaper	Retailer price or discounted if purchased direct from farm	Direct via local retailer or half- or quarter-sides direct from farmer

In the long run, your overall costs of production must be covered in order to have a viable, sustainable business. Enterprise budgets (introduced in chapter 4, "Manage It") are a useful tool for calculating costs of production and establishing a price for your products. Also consider the following key principles:

- Research your market to better understand your competitors' prices and how changes in price affect customer behavior.
- Setting the price too low can communicate that your product is inferior in some way and leave customers wondering if the product is not as good as other vendors' products priced at higher levels.
- Offering bulk discount can be an effective way to move surpluses.
- Generally, direct market outlets provide the opportunity for you to set your prices, whereas in wholesale markets, the buyer tends to define prices based on market prices; in this case, there may be limited room to negotiate.
- Special packaging or value-added processing can help command a higher price.

Published reports and online databases about commodity prices around the country, such as the USDA AMS Organic Prices and USDA AMS Wholesale Market Price Reports (see Resources), can help in your market research. More information on pricing and enterprise budgets is provided in chapter 4, "Manage It."

MARKETPLACE CHALLENGES

The Four Ps and SIVA address factors over which you have some control. In the marketplace, however, there are many factors that you cannot control. These may present some significant challenges:

- **Customers:** Needs, tastes, and preferences change. It is important to know your customers and anticipate changes.
- **Demographics:** Knowing your customers involves learning about demographic characteristics and trends in your region, including age, educational level, ethnicity, cultural background, income level, and generational differences.
- **Competition:** Direct competition is when farmers sell similar products within the same marketplace. Indirect competition involves products that, although dissimilar, can serve as substitutes for one another. Be ready to modify your marketing strategies in response to challenges from competition.

- **Regulations:** Many marketing activities are regulated, including product quality, weights and measures, packaging, and processing. Make sure you are aware of all the regulations that pertain to your products and marketing strategy.
- **Technology:** It can be a challenge to keep up with the pace of technological development, but it is important to stay current in order to take advantage of new opportunities for distribution and payment, and ways to communicate with customers.

A CLOSER LOOK AT A CUSTOMER FOCUS

Direct markets offer small farmers the opportunity to make their business profitable. In direct marketing, a customer focus is essential. Today, food is not just sustenance: for many people, food has become a "values-based" purchase. Food purchasing decisions can be about:

- Leisure time and pleasure
- Entertainment
- Improving health
- Enhancing personal relationships
- Building community
- The environment
- The local economy
- Making a statement about oneself (acting on what you believe in)

These new perspectives on food create new market niches for farmers, where the purchasing experience can be as important as the product itself. This includes shopping at a farmers' market, visiting a farm to pick up a CSA box, picking fruits and vegetables as a family activity, or being on a first-name basis with a farmer. An increasing number of people are looking for a stronger connection to our country's agrarian roots.

One indication that ideas and preferences have shifted is the proliferation of food labels and certifications that communicate certain values and assure consumers that farm products meet certain standards. The table below features several examples of certifications you can find in the marketplace. Take a look at the examples and think about the values that are being communicated by each label. We provide some suggestions, but there are others.

FOOD LABELS AND CERTIFICATION STANDARDS

CERTIFICATION STANDARD	VALUES BEING COMMUNICATED
	Ecologically based production, food safety, health benefits
	Overall sustainability goals
	Humane treatment of animals
	Environmental health, watershed conservation
	Consumer awareness, informed choice
	Social justice, worker benefits

Farmers Share Their Experience
OVERCOMING MARKETING CHALLENGES

THE FOUR Ps (OR SIVA) might lead you to believe that marketing is a simple and straightforward task: just get those figured out and you will be set. But the reality is that marketing is one of the most challenging aspects of running a farm business. Here, Renard Turner discusses how he markets his meat products. Note the challenges he faces and the costs associated with marketing.

RENARD TURNER, VANGUARD RANCH (profile on page 132)

My position as a farmer is I seek to cut out the middleman. You have to learn how to market if you're going to be a small farmer. It's a constant challenge for all of us. I'm a value-added producer, and I cannot stress this enough to anyone thinking about farming: find a niche market, then find a way to become a value-added producer. I've done that by direct marketing my product as ready to eat. I raise a very high-quality herd of myotonic goats and then I turn that meat into goat burgers, curry goat, and goat kebabs that I sell from a food truck. That way I'm able to command my market price.

My story began when I went to the state fair and had a conversation with a vendor, who indicated that selling food through a concession trailer could be profitable. I went home and started looking up concession trailers. Six months later, we had one. And then I had to buy a big truck to pull it.

My unit has everything I need — a refrigerator, freezer, hot and cold running water, a 4-foot grill, and a 2-foot flat grill, plus storage space for spices and utensils. It's 20 feet long and 7 feet wide, and it takes four people to really run it efficiently. It's intense work.

In order to drive traffic to the farm, I built a stage for live music concerts. While people are here enjoying the music, I serve them goat burgers and goat kebabs from the trailer, made entirely from products from the farm — so it's

Renard Turner sells goat burgers and kabobs to concert goers at his farm.

farm-to-table. This way, I don't have to drive up and down the highway and then stand around on corners and try to sell my product.

We grow 20 or 30 different types of all-organic herbs and vegetables, including lettuce, tomatoes, cucumber, squash, and okra. I'm branching out to do some West African vegetables this season so that I can hopefully sell them direct to restaurants and stores in the northern Virginia/D.C. Area. We'll see how that goes. We're always looking for new things to do.

Our marketing is pretty much all by word of mouth. Because we've been farming here for a number of years, local stores know us. I get a call from a store, I go out in the field and take a picture, then I send it to the produce manager and say, "Hey, this is what I have. This is what they look like. They'll be ready to pick in two days. How many would you like? I'll pick them in the morning and deliver them within an hour or two."

We are very high quality. We are organic. We are free-range. I don't compromise. Yes, organic foods are selling at a higher premium, but more people are also producing them. You really have to distinguish yourself by doing something superb and/or unusual or superbly unusual in order to make a niche in this market. There's lots of competition.

What to Market

Figuring out which crops to grow or what livestock to raise can be a complex process based on your vision and values, the physical and human resources of your farm infrastructure, and your economic goals. One step is to consider whether there is demand for a product from wholesale market sources or from direct market sources. Wholesalers and processors generally have specific crops they are looking for. Direct markets are a bit harder to assess.

Using the planning work you did in chapter 1, "Dream It" — your vision, mission, goals, and resources — you can begin making decisions about your products and services. The question of what to grow can actually be rephrased as "Who are my customers (or potential customers), and what do they want?"

To answer those questions, you must do some market research. This could involve researching customer profiles, lifestyle trends related to food (convenience, health, safety, freshness), the ethnic diversity of the population you are reaching and their food preferences, age and income of buyers in your area, and the amount of competition from other farmers. To learn more about how to conduct your own market research, we recommend reading *A Do-It-Yourself Producer's Guide to Conducting Local Market Research* from the Agricultural Marketing Resource Center at Iowa State University (see Resources).

Based on the information you obtain from your market research, think about what products meet the needs and desires of your

ENTERPRISE SELECTION

Take a moment to revisit the Enterprise Selection worksheets you began in chapter 1, "Dream It" (appendices 1-6 through 1-8). Based on what we have covered so far, are there any changes or updates you would like to make? Revisit the sections that are most relevant to your situation and update your answers if needed. Has our discussion of marketing brought any new focus to the type of enterprise you wish to develop?

For additional guidance in this task, read *A Market-Driven Enterprise Screening Guide* from the Western Extension Marketing Committee (see Resources).

STRATEGIES TO ENHANCE PRODUCTS

STRATEGY	EXAMPLES
Service	Added convenience, delivery, portion size, and the like
Quality	Exquisite quality, appearance, and flavor
Production practices	Organic or biological methods, permaculture, low input
Certifications	Third-party certifications for organic production, biodynamic farming, watershed stewardship, or fair labor treatment
Customer experience	Farm visits, farm tours, hayrides, harvest festivals, and so on
Your unique personality	Likability, charisma, other unique personal characteristics that can strengthen customer loyalty

customers and how those products might stand out in the marketplace. Sometimes the product is not currently available locally. Or, if it is, you can look for ways to differentiate it from your competitors. See the table Strategies to Enhance Products above for strategies to highlight and enhance your product.

In addition to helping you decide what products to offer, basic market research can help you figure out how to get customers interested in buying your products and what kind of market niche you might fill.

Offering a unique, quality product may not always be enough to generate sales. You may need to differentiate your product through merchandising — how the products are presented to the customer. The variety of products available and how they are displayed captures the attention (and appetite) of customers and persuades them to make a purchase.

Sometimes a product's uniqueness fills a gap or niche in the marketplace. Niche markets can be profitable if demand for that product is high. However, it is important to remember that a characteristic of niche markets is that they are constantly in flux. Successful niche markets will attract other farmers, thus diminishing their uniqueness over time. The entrepreneurial approach of direct marketing requires being nimble and always looking for new opportunities.

Farmers Share Their Experience

THE FARM AS A FOCAL POINT

THE FOLLOWING INSIGHTS highlight Beth Hoinacki's approach to marketing at Goodfoot Farm. Note what has guided her through the process of developing her marketing strategy — in particular, how her values and goals have helped keep her on track and provided a reference point when making important decisions.

BETH HOINACKI, GOODFOOT FARM (profile on page 44)

An important part of our production system is having a community element — to have our markets be local.

We're about an hour and a half from Portland, and a lot of the growers that I know around here go to Portland markets. I would love to never have to drive farther than one hour to deliver product. So as we develop the farm, we keep an eye on local markets.

Right now, we're doing one Saturday market year-round and an additional market on Wednesdays during the summer. Through farmers' markets, we've been building a list of clients over the years who buy in bulk from us each fall. In mid-October we send out an e-mail to the list saying it's time to place your fall order. Folks will order anywhere from 20 to 100 pounds of potatoes, garlic, garlic braids, and winter squash. We pack all that up for them and do a one-time delivery. That big push moves a lot of product, so we're not storing it and trying to move it all winter. It gives us a break in the winter, which is really nice.

I'm kind of a control freak and somewhat of a perfectionist. The CSA model basically terrifies me because I feel a lot of personal pressure that I need to deliver so many boxes of produce each week and I want it to look a certain way. That being said, I love the community-building aspect of the CSA model. So we are working on developing a CSA focused on members that live in our rural area, and we'll have a weekly box with farm pickup.

When to Market

The third component of your marketing strategy is to define the optimum time for marketing your product: when to market, and over what time frame. Depending on the enterprise, this could mean making your product available throughout the year (for example, meat or dairy) or for just a small window of time (such as fresh fruit in season).

The decisions you make regarding market timing and time frame should align with your cash flow requirements (addressed in chapter 4, "Manage It") and complement the lifestyle choices you identified in chapter 1, "Dream It". Environmental factors that affect the growth and development of your crop (like temperature and rainfall) also influence market timing. Given the variability of environmental factors from year to year, you can control market timing by growing just one crop with a brief or sustained harvest period or by planting a sequence of crops that are harvested in succession to extend time in the marketplace.

You can also control the timing and time frame for marketing through production and processing. For example, season-extension techniques can help you produce a crop earlier and/or later in the season than with standard open-field conditions, as discussed in chapter 2, "Do It." Processing techniques such as canning, freezing, drying, curing, and fermenting increase the shelf life of your product, thereby extending your time frame for marketing. Processing can transform and add value to your product, but you should test the marketplace to determine whether the processed product will sell.

Farmers' markets are the quintessential example of a direct market and offer farmers opportunities to start their business, test products, and expand into additional market channels.

DIRECT MARKET TECHNIQUES FOR SMALL FARMS

Direct markets tend to generate higher profits for small farmers and are easier to access than other marketing channels. The most common direct marketing approaches fall into three categories:

- Farm to consumer
- Farm to retail (including restaurants and food service)
- Online markets

Farm to Consumer

Five of the most common direct-to-consumer marketing channels are farmers' markets, community-supported agriculture (CSA), farm stands, U-pick, and agritourism. As you evaluate the different approaches, refer to your vision and values, your mission statement, and the SWOT analysis you worked on in chapter 1, "Dream It." These are important reference points as you make decisions about your marketing strategy. For additional information on these marketing channels, see the Resources section at the end of this book.

FARMERS' MARKETS

A farmers' market is a place where growers gather to sell farm products directly to consumers. Markets can be located in parking lots, on city streets that are blocked off during market hours, in parks, and in other public spaces. Markets are scheduled for a particular day of the week, and farmers usually have an assigned spot at the market where they display

KEY POINTS ABOUT FARMERS' MARKETS

Farmers' markets offer a secure, regular, and flexible outlet for marketing a wide range and volume of products.

They are a great place for beginning farmers to market their products, and selling at farmers' markets requires low start-up costs.

Farmers' markets provide the opportunity to engage with customers, allowing for direct communication and feedback.

Selling at a farmers' market can lead to other sales, such as restaurants, ethnic markets, or subscription sales.

All farmers' markets have rules that vendors must agree to follow in order to sell at the market; a market manager is responsible for running the market and enforcing rules.

Vendors have to pay a daily fee for the booth space, and some markets also have an annual membership. Many require product liability insurance.

ADVICE FROM OTHER FARMERS

Presentation is extremely important for farmers' market sales; booth design, display, product quality and color, containers, signage, and personal appearance will all contribute to your success.

Farmers' markets can be unpredictable, with wide fluctuations in customer traffic, sales volume, and competition from other vendors.

Other deliveries and errands can be combined with your market day, but this could result in a long and exhausting day off the farm.

and sell their products. Displays range from the back of a farm truck to a simple tabletop to more elaborate covered displays.

At farmers' markets, farmers generally receive retail prices or higher for their products. Because start-up costs are generally low, farmers' markets can provide a low-risk setting for new farmers or an opportunity to try out new products. Many farmers participate in more than one market location to increase their sales. Farmers' markets also provide the opportunity to build a customer base and advertise other outlets for buying the farmer's products. An eye-catching, appealing display with a variety of products is a key factor in promoting sales.

Because start-up costs are generally low, farmers' markets can provide a low-risk setting for new farmers or an opportunity to try out new products.

Farmers Share Their Experience

LISTEN TO YOUR CUSTOMERS

RANDY KIYOKAWA HIGHLIGHTS BELOW a number of the benefits and challenges of farmers' markets and shares some of the specific aspects of his marketing strategy that have made him so successful. Note how he brings a customer focus to almost every aspect of his marketing strategy: What do customers want? What's important to them? What will keep them coming back?

RANDY KIYOKAWA, KIYOKAWA FAMILY ORCHARDS (profile on page 94)

We do a lot more direct marketing now than we did when I was growing up on the farm, back in the '60s and '70s. We used to solely contract out to a packinghouse, and a number of marketplace incidents made me realize that I didn't want to put all my eggs in one basket.

Packinghouses provide a lot of great things, but one of the drawbacks is that you go with the ship. If the sea is rising, a lot of the farms do well, but if there is oversupply, everybody sinks together. I wanted to have a little bit of control over my operation, so we started direct marketing on the farm in addition to contracting with the packinghouse.

Direct marketing started out of necessity with a fruit stand in my garage. We were growing Red Delicious apples but there wasn't a market for them, so growers throughout the country were just leaving them on the tree. I picked a bin of Red Delicious and put out a sign that said 5¢ a pound. If I was working on my tractor and a person came in, I'd wipe off my hands and sell them a box of apples.

Kiyokawa Family Orchards farm stand with playground

Now I have a great crew of local fruit stand employees who are willing to work for a few months of the year in hot or cold or windy or rainy conditions. I find that with the apples, there's demand all the way through Thanksgiving and Christmas.

We belong to an organization that promotes a lot of the local farm stands. There is a brochure with a map and a website. We've been participating in that group for a number of years.

At the peak of our season, we also do about 18 farmers' markets in the Portland area. I lease a barn, cooler, and dock in Portland so that I can stage the fruit there. Satellite shuttles take the fruit to different markets, and the folks who run those markets live in the Portland metro area.

We used to offer tours for tour buses. It's a great way to get started because you're able to learn how to talk to the public and understand your audience. Now I've gotten away from that, just because I don't have time.

You've got to know your customers. Part of that is knowing the heartbeat of what's going on out there, what the trends are. My customers, whether they're at the fruit stand, farmers' markets, restaurants, or stores, tell me what they're looking for and what isn't being produced at that time or in the quantity that is needed, and I plan around that.

"Pick your own" customers at Kiyokawa Family Orchards

COMMUNITY-SUPPORTED AGRICULTURE (CSA)

Community-supported agriculture creates a relationship between farmers and customers. Customers pay a set fee in advance to buy shares for a season. In return, they receive a regular (in most cases weekly) delivery of produce from the farm. Farmers gain financial security by having cash in advance of the growing season and a consistent customer base. Customers get a regular supply of fresh products from the farm and enjoy a sense of community. On some farms, get-togethers with customers or workdays are part of the agreement. Families generally pay for a five- to six-month share. Some CSAs offer winter options or operate year-round.

You need excellent crop management skills to provide attractive and diverse weekly food baskets, as well as good customer service skills. *Approach the CSA model with careful research and planning* so that you can assess consumer interest in your area, and if that interest can translate into the sustained demand needed to make your business successful. Because customers now have so many other options for buying local, it may be useful to think of ways to customize the CSA for your particular niche. There are various types of CSAs, including vegetable, cooperative, dairy, meat, and full diet, and various methods of payment and delivery, such as market bucks, work-trade, pack your own box, delivery, market pickup, and drop sites. You can market your CSA at farmers' markets and through Internet sales.

KEY POINTS ABOUT COMMUNITY-SUPPORTED AGRICULTURE

CSAs require good planning and organizational skills, particularly for setting the planting and harvest calendar and managing the CSA subscription list. If you are filling boxes each week at the same time you are still figuring out your production schedule, you will create a lot of stress.

CSAs offer a secure and reliable market for farmers and healthy and fresh food for customers who might otherwise not have access to farm-fresh products.

CSAs provide up-front income to the farmer at the beginning of the season, which can help with budgeting.

A farm that has an established customer base is in a good position to start a CSA, as current customers can help bring in new members.

ADVICE FROM OTHER FARMERS

Be clear about what is included in the CSA share, and don't give away lots of produce that you could be charging extra for.

Actively engage with customers to help build your customer base.

Provide educational opportunities for members (or prospective members) to come and learn about your farm.

Maintain good communication with your share members and solicit their feedback about the CSA. Many CSA farms produce a newsletter and create other opportunities to interact with members.

Farmers Share Their Experience

CONNECTING THE FARM

ARE YOU RUNNING (or thinking about running) a CSA as part of your marketing strategy? Josh Volk relates his experience with CSA marketing at Slow Hand Farm. He shares a number of important insights about CSAs, in particular how the CSA concept can be adapted and modified to fit your own unique circumstances and to fill a market niche.

JOSH VOLK, SLOW HAND FARM (profile on page 46)

When I first started the farm, I didn't intend to do a CSA. I was thinking that I was just going to do farmers' markets or maybe restaurant sales and do something really flexible where I wasn't completely committed. I had been farming for a long time, but I hadn't been farming on that piece of land.

When I decided to do a CSA, my goal was to do it differently than the CSAs I've worked on. Those had all been family-sized shares with really large quantities of vegetables. One of the biggest complaints was always that there were too many vegetables, and people would leave the CSA because they felt like they were wasting too much of the produce. So I wanted to design a share that was accessible to individuals.

In order to accommodate more people, I broke the CSA year into four smaller seasons. This made the up-front share price a little bit smaller and it also made it accessible to people who only wanted produce in a particular season. Most of the people told me that they liked that the CSA was lower cost and that it had fewer vegetables. I thought of myself sometimes as the gateway CSA for the people who were just trying it out.

The CSA model definitely helps with capital because I got a lot of my cash flow early in the season. I didn't require people to pay up front at the very beginning of the year, and since I broke my season into four distinct pieces, some people were paying at the beginning of each season. Others were just paying in one chunk up front. That combination of up-front working capital and cash flow through the entire season worked really nicely for me.

I think you have to be really careful when you start a CSA. You have to be honest with the CSA members and also honest with yourself about what you're going to be able to provide. In a sense, CSA is the perfect model for somebody starting up because the cash flow is all up front and the community is really supporting what you're doing. But if you're a new CSA, you want to make it clear to your members that they are supporting you as you learn to grow and develop your CSA.

FARM STANDS

Farm stands are structures or displays from which the farm's products are sold. They can range from a simple table on the roadside with a coffee can for collecting money to a dedicated building with refrigerated storage and several employees — much like a store. They tend to be located on the farm, often on a well-traveled road with good access and parking. They can operate seasonally or year-round and may include any number of products. Prices at farm stands are generally similar to retail prices. Depending on the size of your farm stand, you may be subject to local building codes and highway setback regulations.

KEY POINTS ABOUT FARM STANDS

Farm stands are convenient. Selling right at your farm allows for a more flexible schedule since you do not need to make regular deliveries to retailers, farmers' markets, or other venues.

Farm stands encourage interaction with your customers.

Building materials, permits, insurance, and other expensive legal obligations can be a barrier to opening a farm stand.

ADVICE FROM OTHER FARMERS

Increase the diversity of your product line to attract customers and increase overall sales. If needed, you can source additional products from other farmers and think about including specialty products such as jams, condiments or honey, and nonfood items such as artwork, soaps, and lotions.

Make your farm stand a destination spot for customers. Situating the stand near major roads and having appropriate signage will increase customer traffic.

Make sure you provide easy access and ample parking.

U-PICK

U-pick operations invite customers onto the farm to harvest produce fresh from the field. The task of picking the crop — one of the higher costs of growing fruits and vegetables — is passed on to customers. U-pick farms have traditionally appealed to families who do home canning or freezing. Others come to the farm to be in a beautiful setting, have the experience of seeing where their food comes from, and enjoy produce at its peak ripeness.

KEY POINTS ABOUT U-PICK

U-pick simplifies some aspects of marketing, allowing the farmer to focus on growing the product.

U-pick farms often develop a base of committed customers who return year after year.

U-pick reduces labor and post-harvest handling costs.

U-pick offers customers the opportunity to experience farming.

ADVICE FROM OTHER FARMERS

Create an atmosphere that is comfortable and makes picking enjoyable. It may help to have set hours when families with children can come and harvest.

Make your U-pick farm a destination spot. Good advertising is important to bring customers to your farm.

You may need to instruct or supervise customers to make sure harvesting is done correctly.

Bringing people onto your farm property creates potential liability issues. Manage your U-pick operation to reduce risk, and make sure you have proper insurance.

AGRITOURISM

Many people are looking for a stronger connection to the land, want to know more about where their food comes from, and would like to experience what life is like on a farm or ranch. In response to this growing interest, farmers may want to develop an agritourism business. There is a variety of approaches, including:

- **Seasonal:** Pumpkin patch, corn maze, hayride, Christmas trees.
- **Bed-and-breakfast:** An on-farm bed-and-breakfast allows customers to relax in a bucolic atmosphere or actually work on the farm. The concept of farm stays, popular in Europe, is catching on in the United States. For more information, see Farm Stay USA (see Resources).
- **Event focus:** You can host a special farm event, offer a meal or tastings where music and dancing can be enjoyed, or develop an on-farm restaurant. Examples include "farm-to-fork" gatherings or events like those listed at Outstanding in the Field (see Resources).
- **Historical focus:** Other activities, such as cattle drives, attract customers who are willing to pay to experience a celebrated part of our past and present.

KEY POINTS ABOUT AGRITOURISM

Agritourism appeals to a particular type of customer who is looking for the experience of being on a farm and learning about the work involved.

An agritourism business can build community ties and increase the visibility of your farm and farm products.

Returns from agritourism can be highly unpredictable throughout the year; agritourism can also interfere with regular farm activities.

ADVICE FROM OTHER FARMERS

To develop a successful agritourism business, you must have excellent people skills and have an attractive farm for people to come to.

Bringing people onto your farm property creates potential liability issues. Manage your agritourism venture to reduce risk, and make sure you have proper insurance.

Make sure you meet the zoning and licensing requirements in your local area.

Farm to Retail

Many farmers sell directly to restaurants and food retailers. Farmers provide the specific products and high quality that these businesses are looking for, and managers actively cultivate relationships with farmers, often noting the farm name and its products on their menus and in their advertising. Supporting local farms is a philosophical goal for these businesses, and it brings in customers who are looking for high-quality food that is locally produced. You can find similar opportunities for direct sales through the food service divisions of schools, hospitals, and retirement and nursing facilities. In working with these types of buyers, remember:

- Establishing relationships takes time
- Provide samples or tastings of your products
- Establish a process whereby customers can provide feedback about your products
- Quality, consistency, and postharvest handling are important
- Encourage buyers to market the direct farm connection; this can broaden your customer base and educate consumers

Note that restaurants and grocery stores provide an opportunity to develop specialty and niche products, but they often buy in limited quantities, so you will need to set up efficient delivery schedules to maintain profit. It may be difficult to break into the institutional and food service market, so make it easy for them to buy from you. Also, institutions may have special requirements for third-party certification and liability insurance.

Online Marketing and Online Sales

The Internet provides one way to expand the market for your products. An estimated 80 percent of households in the United States now have broadband Internet access, but how do you make yourself known to the people who are most likely to purchase your product? An important first step is to create a website, blog, or social media presence for your farm business, which can help you advertise your farm, sell products, and communicate with customers. A website can also enhance the personal aspect of direct marketing that appeals to so many people. For example, you can use a website to keep customers informed about what is in season and how and where to purchase your products. You can also stay in touch with customers by writing about on-farm or family activities.

It takes some skill to design and develop a quality website, so if you don't have that expertise, hire someone who can help you with that task; it will pay off in the long run. In addition, you will need to register and purchase a domain name and find a web hosting service where you can post your website. For more information on web hosting services, check with your Internet service provider or search online.

If you do not want to create a website for your farm, you can still take advantage of the Internet to expand your market and communicate with customers by creating an e-mail list. Or use the variety of social media platforms available that can provide customers news about product availability and farm events, and provide an opportunity to comment. You can send out regular updates on product availability, farm news, and other information that may be of interest. You might also consider advertising your products on other Internet sites or participating in Web-based farm directories that direct people to your business. Local farm directories are available in most states; a national listing is maintained by Local Harvest (see Resources).

Upgrading your website to provide online sales requires a low capital investment and may enhance market diversification and provide an opportunity for isolated farms to access markets, but it may not be right for someone just starting out. It is much easier to sell product online when you have a broad existing customer base. This type of marketing strategy is built on repeat business and can take time to generate substantial income. Also note the up-front costs (and ongoing labor costs) involved in developing a secure system for online orders, payment processing, and shipping your product. If your online business is still in the early stages, we recommend you conduct in-depth market research to more effectively direct your advertising and sales where they will have the most impact. Numerous books and websites can guide you through the details of building a successful online business. For a quick overview, read *How to Start a Successful E-Commerce Business – 6 Tips from Seasoned Pros* from the U.S. Small Business Administration (see Resources).

Food Hubs

Products that are sold online are traditionally processed so they can be easily packaged and shipped or picked up at the farm or drop site without negatively impacting quality. However, there are now programs and Web-based clearinghouses that provide opportunities to market, sell, and source *fresh* produce online. Food hubs are one example of this mechanism.

A food hub is a middleperson that facilitates the aggregation, distribution, and marketing of source-identified food products primarily from local and regional producers. A food hub is managed by an entity or organization with the primary purpose of connecting producers with diverse markets (wholesalers, retailers, processors, and consumers). There are many food hubs operating in different areas of the country. A national list of food hubs is available from the National Good Food Network's Food Hub Center (see Resources).

...

As you navigate marketing, the goal is to develop a strategy that fits with your vision, mission, and goals and that you feel confident can generate the sales you need. This may involve one or a combination of marketing channels. You may also add new direct market channels as the business grows. For

MARKETING STRATEGY COMPARISON

You can use the Marketing Strategy Comparison worksheet (appendix 1-18) to compare the different marketing strategies covered in this chapter. Choose several of the strategies you are considering for your farm, and write some brief notes on the advantages and disadvantages of each.

instance, you might begin by selling through a farmers' market or a roadside stand, then add other direct channels such as a CSA, or grocery or restaurant sales. The key is to be flexible and adaptable.

MARKET PLANNING

Market planning is the process used by a business to evaluate marketing options and market potential and to develop a marketing strategy. For beginning farmers, market planning generally falls under three categories: jumping in (little planning), enterprise evaluation (some forward thinking and planning), and marketing plan (full planning process).

Jumping in. Many small farmers begin marketing without a formal plan. They grow products they believe have the potential for sales, select a market channel, and simply jump into business. Does this work? The fact that so many successful farms have entered business this way demonstrates that it does. However, perhaps the reason it works is that farmers learn lessons that lead them to do more and better planning.

This approach works well:

- When starting small
- As a method to test the marketplace
- When the risk of financial failure is low
- When the operation does not require a business plan in order to obtain an operating loan

Enterprise evaluation. With this approach, farmers use sales information (or projected sales estimates) to assess their product lineup with varying degrees of specificity. Such assessments include:

- Which products did and did not sell?
- How much of each product did or did not sell?
- What product(s) can be added to improve the current lineup?
- What products(s) must be added to the lineup?
- Which products are not necessarily profitable but necessary to retain customers?
- What quantity should be produced of each product?
- What price is needed to cover production costs and generate some profit?

Evaluating enterprises in this manner is an ongoing task. Every farm business should routinely evaluate each crop (enterprise) for its marketability and profitability.

Farmers Share Their Experience

ADAPTING STRATEGIES FOR MARKET SUCCESS

WHERE ARE YOU along the market planning continuum? These comments from Blue Fox Farm will give you another perspective on market planning. Note the diversity of markets that Melanie Kuegler and Chris Jagger are involved in. And more importantly, note how their marketing strategies have evolved over time, and what has been the driving force for those changes.

CHRIS JAGGER AND MELANIE KUEGLER, BLUE FOX FARM (profile on page 168)

Chris: We sell our produce in a variety of ways — through farmers' markets, wholesale to grocery stores, selling to some restaurants, and then just recently we started selling to a wholesale distributor, Organically Grown Company. We used to do a CSA, but then we switched to a Market Bucks program, which is like a hybrid CSA.

How we market our products has evolved over the years. When we first started out, I remember our original goal was that we were going to be a CSA-only farm. We were only going to grow for this many families, done deal.

We learned several things. We live in a pretty rural area, so if we were just going to do a CSA, we were going to have to always be driving to get to all of our box customers. In the summertime, a lot of people travel and they were away from home, so they missed a lot of boxes. Instead of trying to fix that situation, we looked to a hybrid system.

We came up with the Market Bucks program, which is basically the same idea as a CSA. The customer pays a discounted price in advance for the season (for instance, they pay $425 for a full share, which is $500 worth of produce). This gives us the capital we need to work with at the beginning of the season. In return they get ten $50 cards to use at the farmers' market throughout the season, which is great because they can pick what they want and they can get as little or as much as they want. If they're out of town, they don't have to feel like they're missing their box that week.

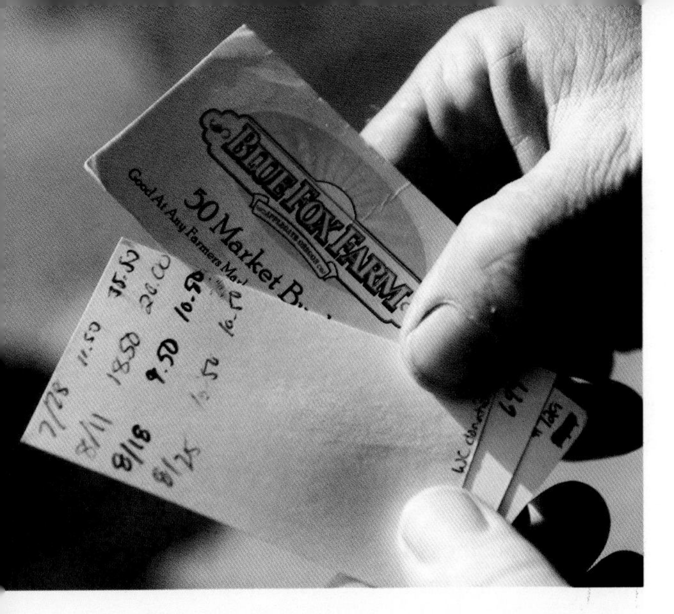

Melanie: The Market Bucks program gives us a little bit of income right at the point in the season where we're having to pay more employees and we don't have any real income to speak of yet. It's just a little bit of a boost for us, so we don't have to rely on credit to get us started for the year.

Chris: We don't grow something unless we know that we can sell it. We've always communicated with our different buyers to see what it is that they really want. You can't always get a firm picture of what people want, but you get an idea. And as we've been here longer and we've seen our customer base stabilize, we know what to expect.

Melanie: We're responding to what the markets want, so what and how much we're growing has definitely evolved over the years. In terms of local sales, we have had to respond to competition from other farms at farmers' markets. That's part of why our wholesale has increased — because of local competition.

Marketing plan. A marketing plan is usually part of a larger business plan. It should be used when starting a new business or introducing a new product. We recommend creating a basic marketing plan as a starting point, even if it is very informal. Answer these key questions: Who are your customers? What crops will you grow? How will you market the crops? What will you charge? What is the competition?

A more formal plan requires extra work, but it can provide better focus and save time and money in the long run. It may involve the entire farm team or an outside consultant to help with the process. A formal marketing plan:

- Assesses the needs of potential customers and identifies products or services to meet those needs (research customers and competition)
- Describes how to communicate the qualities of the product or service to the customer (promotion and advertising)
- Identifies how to get the products/services to the customer (distribution channels)

Although we do not cover it in detail in this book, we highly recommend that you create a general business plan. In fact, as you continue to develop your whole farm plan, you will find that you have already worked on some of the major components of a business plan, including:

- Vision and values
- Mission and goals
- Potential crops/products your farm can produce based on physical resources
- SWOT analysis

The Resources section lists a number of publications, tools, and templates that can help you create a full business plan.

KEYS TO SUCCESS

Marketing your product is just as important and critical to your financial success as growing crops or raising livestock. Listed below are 6 of the 16 "Marketing Musts" from the *Ultimate Small Business Marketing Guide* (see Source Citations) that encompass all the major themes explored in this chapter. Look at each recommendation and its brief description, and think about them as keys to success in developing your farm business.

- **First impressions are vital to business success.** First impressions are lasting impressions. First impressions must carry through all your business and marketing activities. This includes the appearance of your displays, signage, web presence, printed materials, and packaging.
- **Develop a clearly defined, unique selling proposition.** Why will people choose to do business with you instead of your competitors? What one feature or combination of features is going to separate your business from competitors? Better service? Better selection?
- **Know who your customers are and what they need.** The task of knowing your customers and their needs is an ongoing process and requires constant attention. You cannot be complacent. Customers change, their needs change, and the marketplace and competition constantly change.
- **Make it easy for people to do business with you.** Within the environmental constraints of farming, you must be able to provide customers what they want and need, when they want and need it. This means you have to be accessible, and your staff should be knowledgeable about products and services.
- **Make branding the cornerstone of your marketing activities.** Brands sell; they don't have to be sold. Develop a unified image and marketing message and consistently project and deliver it in all your business activities.
- **Never stop investing in yourself.** Never stop investing in ways to make yourself a better businessperson and marketer. Buy and read marketing books and magazines. Attend conferences, workshops, and training courses. Invest in equipment and technology that will help to improve your marketing.

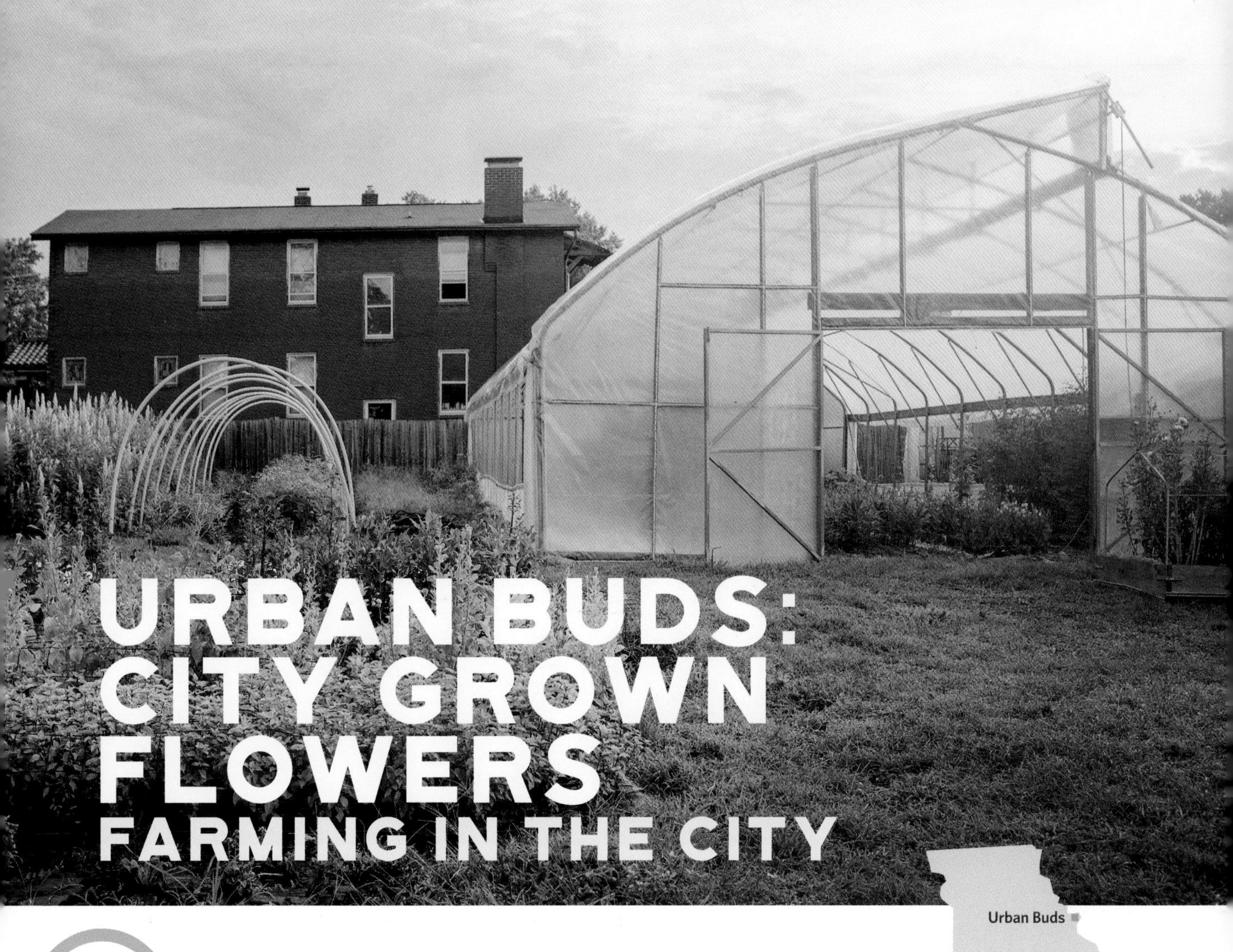

URBAN BUDS: CITY GROWN FLOWERS

FARMING IN THE CITY

Karen "Mimo" Davis and Miranda Duschack have been growing flowers at Urban Buds: City Grown Flowers in St. Louis since 2012. Located just seven miles south of the St. Louis Gateway Arch and less than a mile from the Mississippi River, the farm is a real oasis in the densely populated, residential working-class neighborhood.

Within an eight-block radius of the farm one can find a diversity of businesses, a Catholic boys high school, a mosque, a Southeast Asian Buddhist temple, and the largest Spanish-speaking parish and parochial school in the St. Louis metro area. Mimo and Miranda grow more than 70 different kinds of flowers on the farm, and they sell those flowers locally at farmers' markets and to St. Louis–area florists. They also design floral arrangements for weddings and other events. Mimo and Miranda pride themselves on flowers that are lush, fragrant, unusual, and beautiful.

By growing flowers in this unique urban environment, Mimo and Miranda are connecting with the farm's history. The farm dates from the

At Urban Buds, they are clear that this idea is about more than producing food. It's about building a relationship around something that brings joy to their customers.

1870s, and during the first half of the 1900s it was a vegetable and flower farm on the outskirts of St. Louis. The property diminished over the years as parcels were sold off for development, but one glasshouse and the former florist shop still remain from the original farm.

Urban Buds occupies about one acre in the city. Flowers are grown in the field on raised beds, as well as in a variety of season-extension structures — two greenhouses, a high tunnel, a hoop house, various movable caterpillar tunnels, and low tunnels — that allow them to grow flowers during the winter months with a smaller carbon footprint than imported flowers from other regions.

"Environmental sustainability is a core value," says Miranda. "We use cover crops, compost, and minimal tillage to improve the soil, drip irrigation to conserve water, and integrated pest management techniques to control pests." As part of their urban farm ecosystem, Miranda also manages bees. She is a fourth-generation beekeeper, so her connection to this trade runs deep and is a source of inspiration: "Our bees produce

a rich, fragrant honey that carries with it the many flavors harvested from Urban Buds' flowers." An additional farm business, Urban Buzz, produces honey from the farm.

As they look to the future, Miranda and Mimo are passionate about bringing more educational opportunities and community-building to the farm. They are also developing a new agritourism component to the business, according to Miranda. "Mimo and I love the outdoors — gardening, farming — and we both love teaching. So as we looked at our business model and what makes us happy, we knew we wanted to start being more intentional about the educational component."

At the heart of Urban Buds is the desire to make the world a better place for everyone. For Mimo and Miranda, that core value is borne out in their farm business and in their personal lives as well. Miranda describes how they were one of the first four same-sex couples who successfully challenged Missouri's gay marriage ban in 2014. "We applied for and received a marriage license from the St. Louis City Recorder of Deeds, and we also participated in the subsequent legal challenge to the gay marriage ban that had been approved by voters in 2004.

"When you think about it, isn't organic urban farming a form of activism?" asks Miranda. At Urban Buds, they are clear that this idea is about more than producing food. It's about building a relationship around something that brings joy to their customers.

VANGUARD RANCH
NATURAL GOURMET

Vanguard Ranch, owned by Renard and Chinette Turner, is located in Gordonsville, Virginia, about 60 miles northwest of Richmond. The 94-acre farm business has evolved over the years, and today they focus on raising goats for meat and growing fresh, seasonal produce.

Renard did not come from a farming background. He was raised in a military family and moved around as he was growing up. Then during high school in California, he had the opportunity to join FFA, and that was where his love for farming began. Renard and Chinette lived in Washington, D.C., and eventually decided to move to the rolling hills of central Virginia. "Farming was something that I kind of gravitated to," Renard recalls. "As I began to experience life as a young man in the inner city, it became quite uncomfortable for me to live within that space. I couldn't really see staying there and having to fight through all of the inequalities that were part of everyday existence. Moving to the countryside, and getting out of the city, was escaping all of that to have a better chance." The Turners' initial motivation was to be homesteaders and to become self-sufficient, and those values continue to be at the heart of the farm business. "Self-sufficiency is at the core of all we do. We can feed ourselves. We can keep ourselves warm. Those are two things we never have to worry about. I think that's good."

The Turners also bring a strong ecological and environmental ethic to the farm business. "The environmental aspect is what got me into farming in the first place. I needed to lead a healthier lifestyle, wanting to do my part to make the planet a cleaner, greener place for everyone and generations beyond the one that we're currently living in, doing our part to make planet Earth a better place for everyone."

About 60 acres of the farm is fenced and cross fenced for the goats, and it is managed in a way that mimics the goats' natural habitat. Renard currently has a herd of about 100 goats that he runs on about 25 acres at a time. Initially, he tried raising Kiko and other breeds, but he found that

they could easily jump the fences. So he switched to the myotonic strain of goats because they do not climb or jump like other breeds. Another benefit of this strain is that they have a higher meat-to-bone ratio than other breeds. Renard notes that the low stocking density and the fact that the goats range free in the field are major factors contributing to the health of the herd. Seasonal produce (about 30 different products) is grown on an acre using high tunnels and drip irrigation.

The Turners are true entrepreneurs. From the beginning, their farm business has been about much more than raising livestock or growing crops. A primary goal has been to add value to whatever was produced, and to sell that directly to their customers. "I'm a value-added producer," says Renard, "and I cannot stress this enough: for anyone thinking about farming, you have to become a value-added producer. Find a niche market, then find a way to become a value-added producer."

For the Turners, that means producing antibiotic-free, hormone-free goat meat of the highest quality. Gordonsville is located within a 2-hour drive of several large population centers, so they market the meat direct to restaurants, ethnic markets, and grocery retailers in the region. On top of that, they have a food concession trailer from which they prepare and sell goat kabobs, curry goat, and goat burgers. They bring the concession trailer to county and state fairs, food festivals, music festivals, and other events in the region. They are working on building their own music stage at the farm at which they can host music events and sell their products. The fresh produce grown on the farm (herbs, vegetables, berries) is used for home consumption and for the food concession business, and they also market direct through several food distribution hubs and to area grocery stores.

Renard named his farm Vanguard Ranch because of its reference to being at the forefront and forging new paths. "To be a farmer of color in this country, particularly an African-American farmer, you're going to be faced with a myriad of challenges. They don't seem to be going away, and that's one thing that's for certain. But farming, for me, as an African American, was a way to escape the dependency that our inner cities breed. So you can either stay there and be subjected to the continued politics of America, or you can do something about it." Renard encourages a return to rural livelihoods, especially for African Americans. "What I'm advocating is a paradigm shift — purchasing land, starting small family farms, and transitioning to a rural independence as a heritable way to change the future of your family."

MANAGE IT

GOODFOOT FARM

CHAPTER FOUR

BUSINESS MANAGEMENT FOR THE FARM

As you begin to implement your plans and create your farm business, *business management will be one of the most important factors determining your success.* Whether your interest is in turning a huge profit or making just enough to live on, sound business management is key. Without money to pay for the up-front expenses, it is difficult to get much beyond the planning stage. Without money available through the season for ongoing expenses, you will have a hard time keeping the operation functioning. And without any financial reward at the end of the season, you will be challenged to keep the business going long term.

Business management may seem complicated or intimidating, but with training and practice, all farmers can adopt successful business management practices. Think of business management as a process of continuous improvement. A simplified outline of that process is shown on the next page and includes the following steps:

1. Plan and set financial goals for the business. (You began this process in chapter 1, "Dream It" and chapter 3, "Sell It").
2. Implement plans, grow crops, raise livestock, and create, market, and sell your product.
3. Keep good records of your activities, costs, expenses, sales, income, and other important information.
4. Assess and analyze the results of your efforts.
5. Integrate that information back into the planning process as you build your farm business in the future.

The purpose of this chapter is to help you get started in this learning process.

BUSINESS PLANNING PROCESS

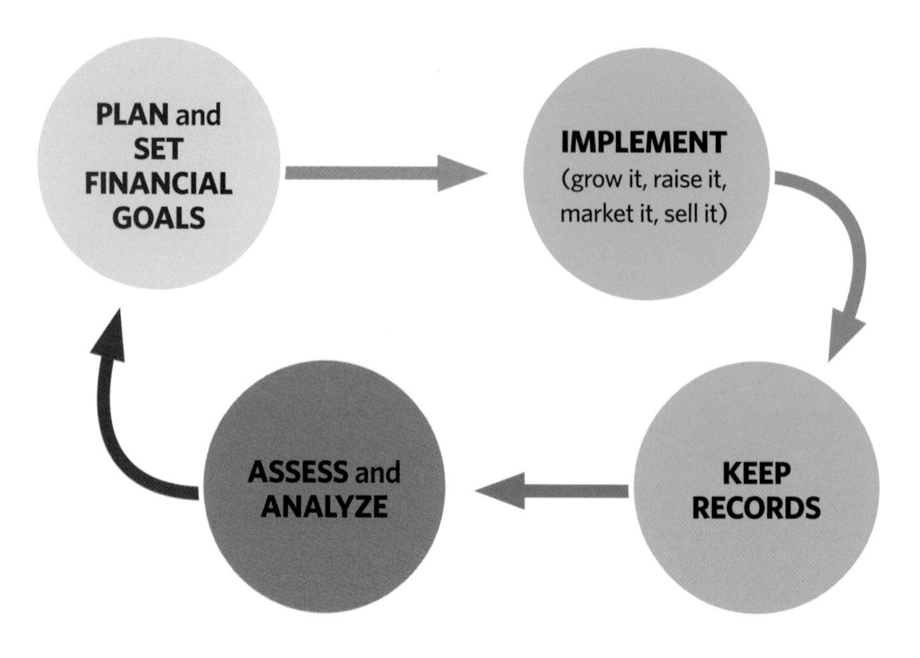

BY THE END OF THIS CHAPTER, YOU WILL:

- Understand the role that different types of budgets have in farm business planning and be able to create budgets for your farm business.
- Know what kinds of records you need to keep track of farm finances.
- Understand the various types of accounting systems and implement an accounting system appropriate for your farm business.
- Be able to create basic financial statements for your farm business that help you monitor income and expenses, manage your cash flow, and assess your financial position.
- Know what kinds of financing options are available to you — and be able to assess those options based on your needs.
- Be able to analyze the differences between what you planned for and what actually happened, and how to use this information to make decisions that allow you to reach your financial goals.

Managing the financial side of your business is a big topic. While we do not cover all aspects, by the end of this chapter you should have a good foundation to work from, as well as ideas for where to look for more information when you want it.

GOAL SETTING AND BUDGETING

Business management begins with establishing some clear financial goals for your farm business. You may want to revisit some of the goals that you identified earlier in chapter 1, "Dream It" — particularly your profit/income goals. Even if you plan to keep an off-farm job in the near term to support your family living expenses while you grow the business, you still must think about how much profit you need each year to cover capital improvements. If your goals include eventually making enough income from the farm business that you can quit your off-farm job, now is the time to get clear on how much you need to make to cover your personal expenses.

Budgeting for Income, Expenses, and Profits

Once you have clearly defined your financial goals, you can look at what you are planning to sell, what that product is going to cost to produce, and what price you can charge in the market. This information will help you determine if and how you will be able to meet your financial goals. To reach those financial goals, you need a budget. *Simply put, a budget is a plan expressed in dollars; that plan is based on estimates derived from the best information on hand at the time it was created.*

INCOME

Your marketing plan (see chapter 3, "Sell It") should include a sales plan that projects how much of your product you plan to sell, when it will be sold, and approximately how much money those sales will generate. These sales projections are the foundation of your *budget for income* — the money you expect to bring in from the crops/products you sell over the course of the season. At the start of the growing season, it is not possible to be exact about these sales figures; it is okay for the budget for income to be based on your best estimates. Conducting some basic market research will help you improve the accuracy of those estimates. Over time, as you keep track of how much you actually sell of each product, the accuracy of your estimates will improve. Pricing can be particularly hard to estimate.

When you are just getting started, you will rely primarily on market prices to determine what you can sell your products for ("competition-based pricing"). Eventually you will be able to incorporate some information about your own costs of production into your pricing decisions ("cost-based pricing").

The chart below shows a sample sales budget for a small farm selling through three different channels: CSA, farmers' markets, and direct to restaurants. CSA sales are projected based on the number of shares sold per week. Projected weekly sales of individual crops at the farmers' market and to restaurants are also listed. This budget shows the expected income for each product the farm is planning to sell, along with the total income for the whole farm.

SAMPLE SALES BUDGET

MARKET CHANNEL	UNITS	QUANTITY PER WEEK	PRICE PER UNIT	NUMBER OF WEEKS	INCOME
CSA SALES	Shares	100	$20.00	30	$60,000
Farmers' Market Sales					
Lettuce	Heads	36	$2.00	16	$1,152
Kale	Bunches	24	$2.50	16	$960
Chard	Bunches	24	$2.50	16	$960
Spinach	Bunches	48	$3.00	16	$2,304
Broccoli	Pounds	100	$2.00	16	$3,200
Restaurant Sales					
Salad mix	Pounds	50	$10.00	20	$10,000
Arugula	Pounds	50	$8.00	20	$8,000

TOTAL INCOME: $86,576

COMPUTER SPREADSHEETS

The sample sales budget was originally created on a computer spreadsheet using Microsoft Excel. Depending on how you set up your spreadsheet, you can have the computer calculate incomes for you and have it recalculate when you change scenarios. For example, in the sample sheet, the numbers in the total income column are generated by formulas. If you change any of the numbers in the quantity, price, or weeks column, the total income number updates automatically, so you can instantly see how changes in your sales plan change the total income projections.

EXPENSES

After you have put dollar figures on your sales plans, it's time to identify how much it will cost you to grow the crops or raise the livestock you intend to sell. It is important to distinguish between operating expenses and capital expenses.

Operating expenses refer to costs for services or supplies that are used up within a year — costs that you incur as you implement your production plan. Examples of operating expenses include:

- Labor
- Seeds
- Plants
- Fertilizer
- Livestock feed
- Electricity
- Fuel

Capital expenses are costs for larger, more durable items that you do not expect to use up in one season. When you are first getting started, you will have a number of capital expenses as part of your start-up costs. Although this will not be the only time that you spend money on capital expenses, these costs will probably be much higher in your first years on the farm. Examples of capital expenses for a farm business typically include assets such as:

- Tractors and implements
- Vehicles
- Buildings for storage and/or packing
- Irrigation systems
- Propagation houses and high tunnels

Capital expense assets are typically in use for multiple years. To evaluate the profitability of the business over time, the cost of purchasing the asset needs to be distributed over the years that it is in use. Doing so aligns these costs with the income (revenues) that it helps to generate. If you do not distribute the expense over multiple years and instead treat the entire cost of purchasing an asset as an expense in the year that you purchase it, you will skew profitability — making the business look far *less* profitable than it really is in the first year, and making it look *more* profitable than it really is in subsequent years.

You can account for capital expenses (assets) through accrual or cash accounting. Accrual accounting is based on the matching principle, which requires that income and expenses be in alignment. To achieve this alignment, capital expenses are distributed over their useful economic life using an accounting tool called depreciation. In pure cash accounting, the cost of an asset is expensed in full when the asset is purchased, even though it will help generate income for years to come. You will learn more about these two different systems for accounting in the Record Keeping section of this chapter.

The sample expense budget on page 141 corresponds with the sample sales budget on the opposite page. Projections for annual operating expenses are listed by cost category, along with the expected total expenses in each category for the whole farm for the year.

In the sample expense budget, labor costs (hourly wages and/or salaries of employees) are included in the payroll category. This category includes an estimate for payroll taxes (often 10 to 20 percent above the wage), and, depending on the status of your hired labor, you may need to pay for workers' compensation insurance and unemployment insurance. Any additional benefits for workers that the farm provides,

DEPRECIATION

The process of distributing the cost of an asset over its useful life is referred to as depreciation. Depreciation accounts for the decline in value of the asset over time and provides a more accurate picture of profitability from year to year.

There are a number of ways to calculate depreciation. As a business expense deduction on your taxes, depreciation is regulated by the IRS, and the methods for calculating it are not necessarily the best ways to evaluate business profitability.

For evaluating farm profitability, the simplest approach is to use straight-line depreciation, where the cost of an asset is distributed evenly over its useful life. To calculate straight-line depreciation, divide the original purchase price of an asset by the number of years of its expected useful life. If the asset is expected to have any salvage value at the end of its useful life, this value is first subtracted from the original purchase price. For example, a $2,100 rototiller that is expected to work for 10 years and have a salvage value of $100 would depreciate at $200 per year, so for each of these 10 years, a $200 depreciation expense would show up on the farm's operating budget. This helps you to align the costs of an asset with the revenues it helps to generate.

There are more complicated ways to calculate depreciation that may give a better representation of the annual cost of owning the equipment. If you are interested in learning about these methods, look for information on "economic depreciation" or "book depreciation" methods and not tax accounting methods. (Be sure to consult with a tax accountant to find out the best way to handle depreciation for your taxes.)

There are no hard-and-fast rules for deciding which equipment to calculate depreciation for, but the general concept is that you want to spread out the accounting for large expenses that represent multiple years of use. Tools that cost less than a few hundred dollars are usually expensed in the first year for the sake of simplicity. For example, you would probably not depreciate the cost of a $50 hoe, even if you expect to use it for many years.

such as health care or retirement, would also be included in payroll expenses. Contract labor or services hired out should be accounted for in a separate expense line. The business structure you choose for your farm (we will cover this in chapter 6, "Keep It") will determine whether you account for the cost of your own labor as part of your payroll expenses or pay yourself out of your net profits.

Note the depreciation expense of $4,000 in the sample expense budget. Fuel costs and the cost to repair and maintain assets are not included as part of depreciation expenses — they are each their own separate expense category. Likewise, the cost of equipment rental is in its own expense category (not under depreciation).

The cost of renting or leasing land is considered an operating expense, as is interest paid on a mortgage, since these are essentially used up in one year. If you buy land, that cost is not treated the same as other capital expenses.

Because land typically holds its value or increases over time and does not have a limited life, the cost to purchase land is not accounted for as an expense at all.

While the sample budget provides a starting point for identifying some of the types of operating costs you need to plan for, it is by no means comprehensive. Nor are we suggesting that the amounts shown in this budget are typical and that they can be used for your farm. Remember, you should plan for the operating costs that correlate with your production plan, and your production plan should be created to achieve your sales plan.

During your first years of farming, you will need to estimate how much labor and other inputs are required to grow or raise the products you plan to sell. As you gather more information and become more experienced, you can update these numbers to better represent your true costs. To prepare an expense budget, you can use these sources:

- Sample budgets (also called enterprise budgets) from educational sources
- Information collected from other farmers
- Seed catalogs
- Quotes from suppliers

Over time, your own actual farm records of costs will likely be the best resource for estimating future costs.

SAMPLE EXPENSE BUDGET

EXPENSE CATEGORY	EXPENSE AMOUNT
Payroll	$46,440
Soil amendments	$4,000
Seed	$3,000
Greenhouse supplies	$1,500
Fuel (on and off farm)	$900
Utilities	$1,450
Repair and maintenance	$500
Equipment rental	$250
Packaging	$150
Market fees	$300
Insurance	$1,000
Land rent	$2,200
Depreciation expense	$4,000
TOTAL EXPENSES	**$65,690**

PROFIT

Once you have budgeted for income and expenses, you can calculate the amount of profit based on those budgets. The sample income, expenses, and profit budget shown below brings together the sample sales budget from page 138 and the sample expense budget at left. The total profit is simply the amount that remains after subtracting your expenses from your income.

Compare this profit to the profit goal that you identified in the beginning of this chapter. If the total profit does not match your profit goal, you need to make some adjustments to your sales plan and your production plan to increase your projected income or decrease your projected expenses — or both.

SAMPLE INCOME, EXPENSES, AND PROFIT BUDGET

Total income	$86,576
Total expenses	$65,690
TOTAL PROFIT	**$20,886**

Farmers Share Their Experience

PRODUCTION COSTS AND PROFITABILITY

PRODUCTION COSTS and the price you get for your product in the market are two major factors in determining your profit margin. As you have learned in this section, there is a real balancing act with both of these factors. Here, Beth Hoinacki and Carla Green share how they are addressing production costs in their own farm businesses. Note the challenges they are dealing with and how each farmer is trying to address those challenges. Note in particular how labor costs figure into the calculation — not just hired labor, but the farmer's own time.

BETH HOINACKI, GOODFOOT FARM (profile on page 44)

I have a time sheet for my employees. When they get to work, they sign in with the time. When we take a break at lunchtime, they sign out for their lunch break. They also write down an approximation of what they did during those hours.

For example, the person who worked for me this summer spent 1½ hours working in the blueberries, prepping the field; 1 hour hilling potatoes; and 1 hour weeding this particular block, maybe the onions. The key thing is that I need to do that for myself, too. I need to write down in my pocket notebook that I just spent an hour wandering around making a to-do list. That's important. That's your management overhead cost, and it has to be accounted for somewhere.

My challenge for this next year is to track my own time and have my husband track his time so that we can establish a crop budget and track production costs. I'm basically setting my prices according to the market, not what it's costing me to produce them. I think you have to take both of those things into consideration. Getting a handle on the production costs is key.

GOODFOOT FARM

CARLA GREEN, SWEET HOME FARMS (profile on page 216)

This is not an easy business to make profitable. You have to look at your costs and your production across the board, and the only way to do that is to keep good records.

When we do our financial planning, we include payment for ourselves to show how much money we're really losing, which is kind of distressing. But it's what you have to do. Once a year, we set aside time to plan for the following year's expected income and expenses. In our projections, we include flock and herd management as well as breeding herd statistics. We use QuickBooks, and it took some time to set those up those accounts, but once set up, they were enormously useful.

For us, labor is a huge overhead cost. We made hay, but we stopped when we realized it cost us as much to make the hay as it cost to pay somebody else to make it. So then we spent that time we would have spent making hay going to farmers' markets to sell, which we felt was a better use of our time.

When we started to do our economic analyses, we discovered that it cost us about twice as much to raise a calf here as it does to raise a calf in central Oregon, so we partnered with a farm in central Oregon. We send our cow calf operation over there, and then we bring the yearlings back here to finish when we have good grass.

Budgeting for Cash Flow

Even if your budget for income, expenses, and profit shows that at the end of the year you will bring in enough income to cover operating expenses — and hopefully have something left over in profit — you could still find yourself short on cash over the course of the growing season.

Cash shortfalls are primarily a matter of timing. In your first several years of farming, you will likely have large cash outflows before seeing any cash inflows. Typically, you will need to buy inputs like seeds, soil amendments, and feed and pay for all the labor required to raise your products before you get a single dollar in sales.

To plan for cash shortfalls, you can create what is called a cash flow budget. A cash flow budget identifies the amount, and the timing, of cash inflows and cash outflows. Cash flow budgets are often developed on a monthly basis but can be created for any time increments that are useful for you.

The key thing to remember is that a cash flow budget is focused on the actual amount of cash going into and out of your business, and the timing of these cash flows. If you are using credit to pay for expenses, account for the cash outflow when that credit is paid for — not when the item was actually purchased. For example, if you are looking at a monthly cash flow budget and you expect to buy fertilizer using a credit card in March but the credit card bill will not come until April, you would record that cash outflow in April.

The cash flow budget at right shows the amount of cash on hand (starting cash) at the beginning of the month and then lists all the cash inflows and cash outflows that occur in that month. At the bottom of each column the cash balance (ending cash) for the month is shown, which is equal to the starting cash plus the cash inflows minus the cash outflows. The ending cash for one month becomes the starting cash for the next month.

In many ways the sample cash flow budget looks similar to the sample income, expenses, and profit budget (page 141). Many of the same income and expenses that show up on the budget for income, expenses, and profit show up here — there is simply more detail here about the timing of this income and these expenses.

A few key differences are worth noting: the sample cash flow budget includes a loan payment and an owner's draw. These are not expenses; they are cash outflows that you need to plan for. A loan payment is not an actual expense — it is a delayed payment for some other expense, typically a way to make the timing of cash flow work better. Any interest on loan payments is an expense. Owner's draws are also not an expense; they are just the owner taking money out of the business, which reduces the owner's equity. Owners also can, and often do, put money into the business (perhaps to buy a piece of equipment). Similarly, that is not considered income — it is a contribution to owner's equity.

Another difference is that cash outflows for capital expenses, such as buying equipment and infrastructure, are included in the cash flow budget. Depreciation, on the other hand, is not included because, although it is an expense, it is not a cash outflow. Remember: in the income, expense, and profit budget, the goal is to compare the income that comes in to the expenses that were required to generate that income — so the cost of assets are distributed over the life of the asset, as a depreciation expenses. In the cash flow budget, the focus is on the actual amount of cash that comes in and out the door — so the full cost of the assets are included as a capital expense.

When your cash flow budget reveals potential cash shortages, you should plan for how to address those shortfalls.

SAMPLE CASH FLOW BUDGET

	JAN	FEB	MAR	APR	MAY	JUNE	JULY	AUG	SEPT	OCT	NOV	DEC
STARTING CASH	**$0**	**$20,993**	**$29,886**	**$33,279**	**$27,172**	**$24,265**	**$23,327**	**$22,889**	**$22,451**	**$22,013**	**$15,906**	**$9,899**
Cash In												
CSA Sales	$30,000	$20,000	$10,000									
Farmers' Market Sales						$2,144	$2,144	$2,144	$2,144			
Restaurant Sales					$3,600	$3,600	$3,600	$3,600	$3,600			
TOTAL CASH IN	**$30,000**	**$20,000**	**$10,000**	**0**	**$3,600**	**$5,744**	**$5,744**	**$5,744**	**$5,744**	**0**	**0**	**0**
Cash Out												
Payroll	$3,870	$3,870	$3,870	$3,870	$3,870	$3,870	$3,870	$3,870	$3,870	$3,870	$3,870	$3,870
Soil Amendments		$4,000										
Seed	$3,000											
Greenhouse Supplies		$1,000	$500									
Fuel		$100	$100	$100	$100	$100	$100	$100	$100	$100		
Utilities	$120	$120	$120	$120	$120	$120	$120	$120	$120	$120	$120	$120
Repairs and Maintenance						$500						
Equipment Rental					$250							
Packaging					$150							
Market Fees						$75	$75	$75	$75			
Insurance	$83	$83	$83	$83	$83	$83	$83	$83	$83	$83	$83	$83
Land Rent	$184	$184	$184	$184	$184	$184	$184	$184	$184	$184	$184	$184
Interest	$50	$50	$50	$50	$50	$50	$50	$50	$50	$50	$50	$50
Loan Payment	$200	$200	$200	$200	$200	$200	$200	$200	$200	$200	$200	$200
Owner's Draw	$1,500	$1,500	$1,500	$1,500	$1,500	$1,500	$1,500	$1,500	$1,500	$1,500	$1,500	$1,500
Capital Expenses												$3,000
TOTAL CASH OUT	**$9,007**	**$11,107**	**$6,607**	**$6,107**	**$6,507**	**$6,682**	**$6,182**	**$6,182**	**$6,182**	**$6,107**	**$6,007**	**$9,007**
Ending Cash	**$20,993**	**$29,886**	**$33,279**	**$27,172**	**$24,265**	**$23,327**	**$22,889**	**$22,451**	**$22,013**	**$15,906**	**$9,899**	**$892**

Farmers Share Their Experience
CASH FLOW ON THE FARM

FARMERS SHARE BELOW HOW THEY MONITOR CASH FLOW and how cash flow challenges have shaped the way they approach the growth and management of their business. What strategies does each farmer use to manage cash flow?

BLUE FOX FARM

MELANIE KUEGLER, BLUE FOX FARM (profile on page 168)

We have felt more comfortable buying newer stuff now because our markets have solidified. In the first few years, we weren't really sure what our year was going to look like in terms of income. Now we have a much more solid picture of that.

When I'm doing the day-to-day bookwork, I'm also looking at our inflow and outflow. I've seen that solidify more, so now we feel more comfortable putting a little bit more into infrastructure and new equipment.

SWEET HOME FARMS

CARLA GREEN, SWEET HOME FARMS (profile on page 216)

We try to butcher almost everything in the fall, freezing it for the rest of the year, so that we have more consistent products and make fewer trips to the processing plant. This is really important, given fuel costs and the labor that goes into transporting.

But this also means a huge amount of money is going out the door for processing that's not going to come back in for many months. We need to plan so that we have the cash flow we need at that time and have enough to cover expenses after that. Processing is one of our biggest costs.

Budgeting for Assets, Equity, and Liabilities

In chapter 2, "Do It," you started to identify the equipment and infrastructure required to get your farm up and running, as well as other assets you will need to acquire over time. You must have access to the capital (money) to purchase these assets when you initially acquire them. For business management purposes, it is important to estimate how much each asset will cost so that you can calculate your total start-up costs (exactly how much money you will need up front). You also must plan for having enough cash on hand (your working capital) to cover your initial operating expenses before any income starts coming in.

After you identify how much your assets cost, you need to address how you plan to pay for them. When you are first starting out and have not yet generated any profits from your business, you need some sources of capital from outside your business. One option is to use your own personal savings to make an equity investment. Another is to borrow money.

When you borrow money to finance the purchase of assets, you are taking on debt and increasing liabilities. You are obligated to repay both the original amount you borrowed (the principal) and the charge for using these funds (interest and fees). Approach this option cautiously and based on realistic financial projections. Lenders look closely at liquidity (your ability to generate sufficient cash to meet financial obligations without interrupting the ongoing farm operation) and the overall financial soundness of the operation.

Especially when you are just starting out, think carefully about where capital funds will come from and the advantages and disadvantages of each source. Below are some of the pros and cons of potential sources of funds for starting your business and for ongoing investments in infrastructure and equipment.

Self-financing. Many farmers start their business with money they have earned through other employment or inherited. Self-financing is a common way to capitalize the farm. In many situations, it may be the only way to develop a business history, which is why detailed record keeping and evidence of improvement in the key business indicators are so important. Self-financing can put personal savings at risk and may compromise family cash flow as the business becomes established.

Family or friends. Borrowing money from family or friends is a viable financing option for many businesses. You may be able to get flexible and attractive terms without needing to put down collateral or fill out complicated application paperwork, like you do for commercial loans. Loan agreements made with family or friends should always be made in writing, with explicit details about the loan terms, including the interest rates, payment schedule, late payment penalties, and default scenarios. Depending on the size of the loan, you should check with a tax professional to make sure the IRS rules are followed. Also keep in mind that borrowing or entering into partnerships with a loved one could strain the relationship if challenges arise with the farm's or lender's finances.

Crowdfunding. There are various forms of crowdfunding, but the basic idea is that a large social network helps fund your project. The funds may take the form of donations, payments for rewards, or microloans. Equity investment from individuals in businesses is regulated by federal and state governments, so these platforms generally don't include the option for funders to own "shares" in the business, but often donations are given some sort of "reward" or interest is earned on loans.

Commercial bank loans. Commercial banks offer various types of credit for small farms. Farm Credit is a nationwide network of borrower-owned lending institutions and specialized service organizations. Local, regional, or national banks may offer farm loans in certain circumstances. Generally, commercial lenders require the following to consider a loan:

1. Three years of financial records that show adequate profits and cash flow
2. Sufficient personal or business collateral (to back the loan in the case of a default)
3. A credit score that indicates that the borrower has a track record of meeting other financial obligations

Commercial banks rarely make loans to start-up farm businesses. Beginning farmers face a high level of scrutiny in securing a loan and must demonstrate a compelling case that the loan makes business sense. Commercial banks have high standards for documenting income and assets, making good accounting records critical.

Government loans. The USDA Farm Service Agency (FSA) offers farm operating and farm ownership loans. These loans are designed for farmers who are unable to obtain private, commercial credit. Both guaranteed loans and direct loans are available through the program. Under the guaranteed loan program, conventional lenders (banks, Farm Credit system institutions, and other lenders) make the loan, and FSA guarantees the loan. Direct loans are made and serviced by FSA officials, who also provide borrowers with supervision and credit counseling. Applicants must show sufficient repayment ability and pledge enough collateral to fully secure the loan. Farmers apply through their local FSA office. As with commercial loans, small farmers who have limited capital and are in their first few years of farming will find it difficult to secure a government loan.

Local economic development agency and Small Business Administration loans. Local economic development agencies are generally structured as nonprofit organizations. Their mission is to help entrepreneurs with low incomes and/or credit issues. Many of these agencies offer loans to businesses that are not able to secure bank financing. Some economic development agencies may also offer other business advising services.

Business funding. Several businesses offer financing at potentially attractive terms. Whole Foods Market's Local Producer Loan Program provides loans for capital expenditure to small-scale farmers. Their website states that they minimize the fees, interest rates, and paperwork. Equipment companies such as Kubota, John Deere, and Massey Ferguson at times offer zero percent financing for up to 60 months on purchases of qualified equipment. Other equipment and agricultural input companies may offer financing on a variety of terms. As with all financing, farmers should carefully scrutinize the terms of the loan and their ability to repay the debt. Zero percent financing from equipment companies may distort financial realities and encourage purchases that are not justified or wise.

Credit cards. Credit cards may be useful for short-term financing, offering similar advantages and disadvantages to using them for personal expenses. Credit cards offer easy-to-obtain and noncollateralized financing for those with good credit scores, and they often offer attractive introductory terms, like low interest or even zero percent interest for one year. Be cautious about using credit cards for farm financing, since interest rates are generally variable and move higher with time. Most introductory interest offers end after a set period — often one year or less. Farmers who are unable to pay off credit cards before the introductory period ends will be stuck with high-interest debt that is difficult to pay off.

In addition, credit limits are often lower than you might need for the business, and a credit card can't often be used for certain types of expenses (such as labor and used equipment purchased from other farmers).

Government subsidies or grants. Federal agencies offer grant funding for farm improvements, conservation, and on-farm research. Two potential sources are the Sustainable Agriculture Research and Education Program (SARE) and the National Resource Conservation Service (NRCS). Grants generally require extensive paperwork and are useful only for select elements of the farm's funding needs. The availability of government grants is dependent on their funding and varies from state to state, despite being a federal program.

Individual development accounts. An individual development account (IDA) is a matched savings account that enables limited-income families to save and build assets. IDA participants save a set amount of money; upon the successful completion of the savings terms, they receive a grant of matching funds (often two to four times the amount saved). These programs receive funding from a variety of private and public sources and are frequently administered by nonprofit organizations. These programs are often subject to strict rules regarding the savings period and amount and the types of purchases the funds can be used for. IDA program participants are often required to complete business planning and business education classes, which can offer additional value to the participant.

. . .

The table below shows a sample budget for assets, liabilities, and equity. On the left side are the farm's assets and the value of each asset. On the right side of the budget are the amounts of equity (self-financing) and borrowed money (liabilities) that were used to acquire these assets. Notice that the total assets are equal to the combined total of equity and liabilities — they are in balance. This budget for assets, liabilities, and equity mirrors the balance sheet, which you will learn about later in this chapter (see page 156).

SAMPLE BUDGET FOR ASSETS, LIABILITIES, AND EQUITY

ASSETS		EQUITY	
Cash	$10,000	Personal savings	$40,000
Tillage tractor	$15,000	**LIABILITIES**	
Cultivation tractor	$25,000	Bank loan	$20,000
Greenhouse	$10,000	**TOTAL EQUITY + LIABILITIES**	**$60,000**
TOTAL ASSETS	**$60,000**		

There will always be places where you are still guessing, but as your experience grows, you will benefit from your prior work in creating budgets and keeping good records, leading to better guesses in the future.

Keep in mind that there is no one right way to finance the costs of starting up your business. It depends largely on your personal or family financial situation and your comfort with taking on debt. It is important that you have a plan for how you will obtain the capital necessary to get started making money. Over time, any profits that you make can be used to pay down debt and/or can be reinvested in the business (used to acquire additional equipment and infrastructure).

. . .

Ultimately, the accuracy of a budget is only as good as the quality of the information used to create it. When first starting out, you must rely primarily on information from outside sources and your best guesses. As the farm matures, you will have your own records to look at and your budgeting will be more informed. There will always be places where you are still guessing, but as your experience grows, you will benefit from your prior work in creating budgets and keeping good records, leading to better guesses in the future.

RECORD KEEPING

The first step in the business management process is to clarify your financial goals, and the second step is to plan for how you will reach these goals. The three types of budgets highlighted in the previous section are a key element in that planning process. As you begin to implement your plan, you quickly realize that things do not always go as expected. Keeping good records will help you learn what is actually happening, refine your plans, and continuously improve your farm business.

Accounting Systems

To keep records on the financial aspects of your farm, you must set up an accounting system that tracks income, expenses, profits, cash flows, assets, liabilities, and equity. A useful accounting system has the following characteristics:

1. It is easy to input your data and retrieve that data for the purposes of analyzing your finances.
2. The system helps you maintain the appropriate records for filing your local, state, and federal taxes.
3. Unless you receive immediate payment for everything you sell, the system should allow you to create invoices and keep track of payments so that you can easily tell who has paid you and who still owes you money.
4. If you have employees, you need a system that allows you to run payroll.

Be aware that any accounting system that meets all of the above requirements is going to be somewhat complex and may not feel easy and intuitive the first time you use it. A comprehensive system is going to take some time to learn, so you should make sure to allow both time and funds to pick out and learn a system that matches your needs and helps you grow as a business. The good news is that you do not need to create a system from scratch. There are many good computer-based systems out there today, and even paper-based systems are still in use and can be quite robust.

HIRING AN ACCOUNTANT OR BOOKKEEPER

Professional accountants can help you set up and maintain an accounting system for your farm business — one that allows you to keep good records and easily retrieve that information for later analysis. A good accountant usually has a degree in accounting and understands good accounting practices. A certified public accountant (CPA) is an accountant who is certified by the state to prepare and file your taxes.

Bookkeepers generally have less training than accountants, but a good bookkeeper may be able to perform many of the same tasks when it comes to setting up and maintaining an accounting system. Because bookkeepers usually have less training than accountants, they can be less expensive to hire. A good bookkeeper can also save you money by reducing the amount of time it takes a CPA to prepare your taxes, but the bookkeeper will not be able to prepare your taxes for you.

Many farmers (especially those just starting out) learn to do their own bookkeeping, and some even do their own taxes. Yes, hiring a bookkeeper and/or a CPA is an added expense, but it helps you keep your books accurate and up-to-date. This in turn can ensure that your taxes are filed appropriately, and on time, and that you are paying only the taxes that are required, saving you money in the long run. If you do not like meticulously recording numbers, you might be better off hiring a bookkeeper.

When you hire a professional bookkeeper or accountant, he or she should help you set up your accounting system so that you understand how it works. Be aware that most accountants and bookkeepers are focused primarily on tax issues, whereas you, the business manager, also want to make sure your accounting system tracks the areas of your business you wish to analyze. It is generally best to set up your accounting categories in a way that makes sense when you are analyzing the farm business, then let the accountant sort out what tax category the accounting categories go into at tax time.

CHOOSING AN ACCOUNTING SYSTEM

Setting up a good accounting system from the start will save you a lot of time down the road. By making the information you need more accessible, you will facilitate business planning and analysis. Here are a couple of questions to consider when setting up your accounting system.

Which accounting method should I use — cash or accrual basis? Once you have chosen an accounting method, it is not easy (from a tax perspective) to switch, so consider this decision carefully.

With accrual-basis accounting, income is recorded when it is earned. For example, when a sale is made and the invoice is created, it is considered income — even if the payment has not been received. Likewise, expenses are recorded when they are incurred: that is, as soon as the bill comes due, whether you have actually paid for it or not. Accrual-basis accounting provides a more accurate picture of profitability because it is better at aligning income with the expenses incurred to generate that income. The downside is that it can be more complicated and expensive to implement, and it is often not a feasible option for smaller farms. Certain farm corporations and partnerships should (or are required to) use an accrual method of accounting.

With cash-basis accounting, income is recorded when payments are received and expenses are recorded when payments are made. Although cash accounting is simpler to use and can provide certain advantages in terms of managing tax liability, the cash basis can skew profitability. For example, say you sell a large amount of product in December of this year but you don't get paid for it until January of next year. If you are on a cash basis, it will look like you brought in less income this year than if you are on accrual basis. If the expenses that are associated with these sales were recorded this year, then your profits for this year will appear lower than they actually are.

One option that allows for using a cash-based accounting system but still provides a more accurate perspective on profitability is to make what are called "accrual adjustments." Accrual adjustments are basic additions and subtractions made to a cash-based income statement. You can learn more about making accrual adjustments from *Converting Cash to Accrual Net Farm Income* and *Your Farm Income Statement* and *Preparing Agricultural Financial Statements* (see Resources).

What tax rules and options apply to my farm business? Farmers should carefully study the *Farmer's Tax Guide* published by the Internal Revenue Service (IRS) (see Resources). This free online publication explains the tax rules for farm businesses and the various options for tracking and recording financial information. One choice that farmers need to make is deciding which method of depreciation for durable property to use. The IRS allows for the General Depreciation Schedule (GDS) and the Alternative Depreciation Schedule (ADS). GDS is popular because it has a shorter recovery period (number of years in the depreciation schedule). This allows for greater deductions on your income tax in the short term. However, for farmers wishing to show a greater net worth (such as in the case of applying for loans), the ADS shows a higher value for the property because the recovery period is longer. (See *Farmer's Tax Guide* "Table 7-1: Farm Property Recovery Periods.") We recommend consulting with a tax professional to establish the best position with regard to your tax bill, while ensuring that you are in compliance with state, local, and federal tax regulations.

Are commercial accounting products worth considering? One of the advantages of using accounting software designed for businesses is that in learning how to use the software, you also learn good accounting practices. It is easy to develop bad accounting habits, so it is often worth reading accompanying manuals or working with a professional to set up the system so that you do not run into problems later.

If you hire an outside bookkeeper or accountant, you will most likely adopt the accounting system (and software) used by that individual. Below are three commercially available accounting systems you might opt to use.

- **Manually written ledgers:** Paper is still an option, and you can buy books and ledger sheets specifically designed to keep manual records.
- **Computer spreadsheets:** Common software like Microsoft Excel, Numbers, Google Sheets, and OpenOffice Calc are all excellent tools for budgeting, record keeping, and basic financial analysis.
- **Small business accounting software:** An ever-growing list of online and desktop versions of software and services like QuickBooks, Sage 50, and GnuCash can organize business records, including invoicing, bills, and payroll, and automate many parts of the bookkeeping.

Most farmers use a combination of all three options, depending on what records they are keeping, what they are analyzing, and if they are using an accountant or bookkeeper. Each has its own place on the small farm.

Keeping Good Records

Accounting systems require good organization and thorough documentation. Make sure to properly back up your records off–site or online in case paper records are destroyed. At a minimum:

- **Maintain paper and electronic copies of all sales invoices and payments received.** Keep a record of the date the product was ordered, who ordered it, when it was delivered, and when it was paid for (including payment information such as the check number).
- **Maintain paper and electronic copies of receipts** for all purchases.
- **Collect signed sales receipts for all orders.** Deliveries to the customer should be verified by obtaining a signed receipt by an authorized recipient. Receipts or invoices for sales should list the payment terms of that sale (for example, payment due upon receipt or due in 30 days). Issue invoices for payment due in a timely manner; many farmers find the best way to issue invoices is at the time the product is delivered in the form of the delivery receipt (the delivery receipt and invoice are the same).
- **Keep a cash record.** Many farmers collect large amounts of cash from farmers' market sales; it is important to track what cash is received and where it is deposited.
- **Stipulate the terms of sales** for restaurant and community-supported agriculture customers. Terms should be clearly stated to customers in writing, including due dates and late charges. Community-supported agriculture businesses should ensure that online or paper signups include details about the membership terms, including refunds, vacations, or missed pickups.
- **Document labor.** Keep time cards for the farm operator and employees. In many areas, there are different insurance rates for agricultural labor, but on small farms some tasks do not fall into the category of agricultural labor so you may need to track different tasks on time cards separately. You can also use categories on time cards to track different enterprises or tasks on the farm. For example, if you have employees who staff a farm stand and also work in the field, you may decide to track the hours they spend working the farm stand separately on time sheets to help analyze the farm stand enterprise.

Farmers Share Their Experience

RECORD KEEPING: FINANCES AND INVENTORY MANAGEMENT

WE HAVE EMPHASIZED the importance of keeping accurate records — for assessing the financial state of the farm, for tax purposes, and for planning the future of the farm business. The farmers quoted here reinforce this message. As you read their comments, note the kinds of data they are tracking and the methods they use to collect, record, and manage that information.

BETH HOINACKI, GOODFOOT FARM (profile on page 44)

My bank statements are key, so I maintain a separate bank account for the farm. It's the simplest thing in the world to set up: it's just a second checking account with the farm name on it.

Both my husband and I have debit cards for the farm account, so if we're buying something for the farm at the feed store, we try to remember to use that card. That way, we can keep track of costs for the farm with our bank statements and our receipts.

GOODFOOT FARM

I stuff every single farm receipt into one file in my desk. Now, if I was really good about it, I would divide the file into categories, but at least the receipts get into the folder. When tax time comes, I simply pull out all of those receipts.

I got an accountant because we were doing a lot of capital investment in infrastructure to get the farm going, and she helped with depreciating things. I do payroll taxes myself, but I call her up and run things by her. She answers my questions and gives me advice. She's a really valuable resource for the farm.

At the farmers' market, I track my sales in a notebook. It's kind of laborious and it's not going to work forever, but it works for me right now because I'm not selling a huge amount of different things.

I write down weights, if it's a scale item, or the number of bunches and what the produce is. I have another sheet with prices, so I can look back and calculate how many pounds of potatoes or bunches of kale I sold at a particular market.

At the end of the year, if I have the time and inclination, theoretically, I could sit down with these notes and calculate exactly how many pounds of potatoes, bunches of kale, or pounds of apples I sold — that sort of a thing. And hopefully, that could link back to some production costs.

RANDY KIYOKAWA, KIYOKAWA FAMILY ORCHARDS (profile on page 94)

I use an accountant for year-end record keeping, and I have a bookkeeper who comes in once a month. I like to pay my own bills, and I like to do my own payroll. That way, I keep a pulse on everything that's going on.

I use QuickBooks. We've tailored it to fit our operation, and the fact that we pay both hourly and sometimes contract. I feel like I can make my best decisions if I know what's really going on financially with the operation.

When I do the inventory for a farmers' market, such as how many boxes we're going to send out, I put those numbers on a spreadsheet so that I can track them and can copy and paste to the next week. I do the same thing with the inventory for the restaurants and stores that we deal with.

CARLA GREEN AND MIKE POLEN, SWEET HOME FARMS (profile on page 216)

Carla: We used QuickBooks for our finances. The program requires some understanding of accounting, which we were sorely lacking in the start. It took a while to figure out how to use the program.

We had accountants prepare our taxes. We had our property's LLC [limited liability company] and also the farm LLC, so our taxes were pretty complicated and, unfortunately, expensive to get done. But it was just beyond our capacity.

Fortunately, as we got better with QuickBooks, we provided better records to the accountant. We also had a bookkeeper who did our payroll for us, which was very helpful.

We are trying to come up with a record-keeping system for our inventory management. We don't have a way to track the sale of each cut of meat. We are trying to figure out a way to do this.

Mike: We struggled with being able to reliably predict the yields of salable meat from, say, an annual crop of calves or an annual crop of lambs. In other words, if we started with *x* number of lambs, what is the average hanging weight for those animals and the average meat yield from those weights?

Our yield estimates improved, but we would have liked to be able to estimate how many cuts of various kinds we would have available to sell so that we could have a truer estimate of our income over the next three to six months.

We did a pretty good job, I think, of keeping records for different aspects of the enterprise, but we struggled with finding the time to analyze them and effectively use that information.

FINANCIAL STATEMENTS

A key function of your accounting system is to keep records of what actually happens with farm finances. Financial records are typically organized into reports called financial statements. There are three financial statements commonly used by most businesses: a profit and loss statement, a cash flow statement, and a balance sheet. These statements are closely related to the budgets that we looked at earlier. The difference between a budget and a statement is that a budget is a projection or plan for what you think will happen in the future, and a statement is a record of what actually happened.

Profit and Loss Statements

The profit and loss statement (P&L) shows income that was earned, expenses that were incurred to earn this income, and the profit that remains. The P&L measures a business's net profit over a specific accounting period, such as one month, one quarter, or one year, with one year being the most common period used. The P&L is also known as the "income statement" or "statement of revenue and expenses." The purpose of the P&L is to show whether the business made or lost money during the period being reported. The profit and loss statement mirrors the budget for income, expenses, and profit we looked at earlier. The two can be used together to assess differences between what was projected and what actually occurred.

Cash Flow Statements

Cash flow statements show where cash came from and how cash was used; it explains changes in cash assets from one period to another. The cash flow statement demonstrates a business's ability to cover current expenses, pay debts, and invest in new equipment and infrastructure by highlighting cash surpluses and cash shortages. The cash flow statement is directly related to the cash flow budget that we looked at earlier. Both show cash inflows and outflows over a period of time. Cash flow budgets project what future cash flows are expected to be; cash flow statements document actual cash flows. A cash flow statement differs from a profit and loss statement in that it focuses only on changes in cash and does not indicate the profitability of the business, since some cash inflows are not income and some outflows are not expenses.

Cash flow statements typically categorize cash based on the business function it was used for: operating activities, investing activities, or financing activities. Cash used for operating activities includes cash inflows and outflows related to the core functions of a business. For farmers, core functions are producing and selling products from the farm. Cash inflows and outflows related to purchases or sales of equipment and infrastructure are categorized as investing activities. Financing activities are related to cash inflows and outflows from loans, credit lines, mortgages, and owner's draws.

Balance Sheets

The balance sheet is also known as a "statement of financial position." This statement is a snapshot in time that shows the value of the business's assets and the liabilities and equity that support these assets. More simply, the balance sheet shows how much the business owns and how much the business owes at a moment in time. The value of the farm's assets is always equal to (and "in balance" with) the combined values of the farm's liabilities and owner's equity. Earlier in this chapter, when you budgeted for the assets that you need to purchase and then planned for how to finance these assets, you essentially created a projected balance sheet. Over time, as profits are generated and reinvested in the business — either by buying additional equipment and infrastructure or simply by keeping more cash on

hand — the owner's equity also increases and the farm's financial position improves. These changes are reflected in changes in the balance sheet from one period to the next.

. . .

The financial statements described here are typically requested by lenders, business partners, or investors when they want to make an assessment of your business. They are also useful (in combination with the budgets you have developed) for doing your own assessment and analysis of how well your farm business is meeting your financial goals.

ANALYSIS

The final step in the business management cycle is to use your budgets and records together to analyze your farm business. This will give you a clearer understanding of your farm's financial situation, so that you can make informed decisions about how to revise plans and budgets going forward.

Variances

One of the simplest methods for analyzing your farm business is to look at variances. A variance is the difference between a projected budget amount and the amount reported in your financial statements (what actually happened over the course of the season). If you have a good accounting system in place, you should be able to quickly generate financial statements, and the categories in those statements should match the categories in your budgets in a way that allows you to easily calculate the variances.

To analyze variances, first look at what may have caused the difference. After you have identified the cause, you can evaluate the impact of the variances and how to address any problems. For variances that have a negative impact on profits (such as a shortfall in sales or overspending on labor), you need to come up with solutions for fixing the problems before they become catastrophic. If the variances are in your favor (such as better-than-expected yields and sales or lower-than-expected spending on supplies), you should still do a quick check to make sure the source of those variances is not going to cause problems in the future. Examples that might cause future headaches include future shortfalls in product to sell or problems with long-term fertility in the fields. As long as there are no potential problems, you can pat yourself on thc back for a job well done in creating a conservative budget and exceeding expectations.

Analyzing the Profit and Loss Statement

Variances are particularly useful in analyzing the difference between projected and actual income, expenses, and profit. The primary benefit here is to make future budgets more accurate.

The table below shows a comparison between a budget for income, expenses, and profit and the actual income, expenses, and profit from a profit and loss statement. What variances do you notice in this table?

When evaluating variances, a good starting point is to look at the difference between the amount of projected profit and the actual profit that was earned. If the difference is favorable (actual profits were more than projected profits), you will probably be pleased. An unfavorable difference (actual profits were less than projected profits) will be cause for concern. Both scenarios warrant further investigation to see how well you budgeted for each income and expense line, and to learn from the variances. Here are some

SAMPLE PROFIT AND LOSS STATEMENT SHOWING VARIANCES BETWEEN BUDGETED AMOUNTS AND ACTUAL AMOUNTS

Example of variance from a budget to the actual amounts on the Profit and Loss statement. (U) indicates unfavorable; (F) indicates favorable.

CATEGORY	BUDGETED	ACTUAL	VARIANCE
INCOME			
CSA sales	$60,000	$60,000	$0
Farmers' market sales	$8,576	$9,000	$424 (F)
Restaurant sales	$18,000	$17,500	$500 (U)
TOTAL INCOME	$86,576	$86,500	$76 (U)
EXPENSES			
Payroll	$46,440	$48,000	$1,560 (U)
Soil amendments	$4,000	$4,200	$200 (U)
Seed	$3,000	$2,800	$200 (F)
Greenhouse supplies	$1,500	$1,500	$0
Fuel (on and off farm)	$900	$100	$800 (F)
Utilities	$1,450	$1,450	$0
Repair and maintenance	$500	$800	$300 (U)
Equipment rental	$250	$250	$0
Packaging	$150	$200	$50 (U)
Market fees	$300	$300	$0
Insurance	$1,000	$1,000	$0
Land rent	$2,200	$2,200	$0
Depreciation expense	$4,000	$4,000	$0
TOTAL EXPENSES	$65,690	$66,800	$1,110 (U)
TOTAL PROFIT	$20,886	$19,700	$1,186 (U)

questions you might ask yourself if you did not reach your profit goal:

- Are actual sales less than the sales that were budgeted/planned? If so, why?
- Were yields less than expected?
- Was selling the planned amount of product a challenge?
- Did prices need to be adjusted?
- Were there other challenges in the marketplace?
- Are actual costs more than the costs that were budgeted/planned? If so, which costs were over budget and why?
- Was the cost of labor higher than planned, or did you use more labor than budgeted for?
- Did other inputs (fertilizers, seed, feed, fuel, and so on) cost more than you budgeted or did you use more than you planned?

IMPROVING PROFITABILITY

If you are not meeting your profit goals, you should consider what operational changes can be made to improve profitability. And even if you are profitable, you may want to look at ways to become more profitable.

Increasing sales. Assuming you sell your product for more than it costs to produce, increasing sales will improve your profitability. This may mean doing more market research, including conversations with your customers where possible, to make sure there is a market for what you are producing. Soliciting customer feedback is essential — both before and during the season. If there is a quality issue (or some other issue with your product), it is far better that you find out about it and have the chance to correct it before you lose sales or customers.

Increasing yield. If yields are lower than planned, this will impact both the income you bring in (reducing potential sales) and your costs of production (since preharvest production costs will have to be spread over more units). Strategies for increasing yields include choosing crop varieties and animal breeds that are suited for your location and identifying optimal planting dates and feeding schedules. To increase yields, you may need to increase some of your spending on inputs, such as fertilizer and water. Weed pressure can have a big impact on yield; investing in labor to control weeds often pays off with increased yields and reduced harvest and cleaning time.

Reducing expenses. When it comes to reducing expenses, your financial statements can steer you toward the expense categories that have the biggest impact on total costs. It may help to list all your expenses and calculate what percentage each category contributes to the total. The Expenses as Percentage of Total Expenses table below provides an example. Looking at expenses this way can focus your efforts on the expense category that will have the greatest impact. In the example, note that the payroll expense makes up almost 72 percent of total expenses. This area may have the most room for improving the bottom line (profits).

EXPENSES AS PERCENTAGE OF TOTAL EXPENSES

EXPENSE CATEGORY	EXPENSE AMOUNT	PERCENTAGE OF TOTAL EXPENSE
Payroll	$48,000	71.9%
Soil amendments	$4,200	6.3%
Seed	$2,800	4.2%
Greenhouse supplies	$1,500	2.2%
Fuel (on and off farm)	$100	0.1%
Utilities	$1,450	2.2%
Repair and maintenance	$800	1.2%
Equipment rental	$250	0.4%
Packaging	$200	0.3%
Market fees	$300	0.4%
Insurance	$1,000	1.5%
Land rent	$2,200	3.3%
Depreciation expense	$4,000	6.0%
TOTAL EXPENSES	$66,800	100.0%

Strategies for reducing labor costs fall into three general categories:

1. **Timing.** Focus on getting things done on the farm at the right time. Weeding can take twice as long (or more!) once weeds have gotten away from you — not to mention that weeds that go to seed will cost you more in labor in future seasons.
2. **Training.** Focus on training your employees (and yourself) to work efficiently and ergonomically.
3. **Technology.** Focus on using appropriate technology to reduce labor costs. This does not mean you need to mechanize everything on the farm, but certain investments in equipment and machinery may be worthwhile in terms of the labor expenses they will save you.

In addition to the strategies above, you may want to consider outsourcing functions that you, or someone on your farm team, are not skilled at or do not have the capacity to handle. A new farmer may find it makes more financial sense to hire skilled services for certain tasks. It's better to hire someone on contract when needed than to struggle and waste valuable time trying to get the job done. Tasks that are done incorrectly can produce results that hurt the business, ultimately costing more money.

Another strategy for reducing costs is to buy production inputs in bulk quantities. If you have a good place to store seeds or feed, it can pay off to buy a full season's worth rather than small quantities here and there. Consider pooling orders for supplies with other farmers in order to buy in larger quantities and benefit from bulk discounts.

Finally, consider if it makes more sense to rent equipment rather than purchase it. Note in the sample profit and loss statement on page 158 that there is a depreciation expense of $4,000. Suppose this includes the annual ownership cost of a $25,000 tractor distributed over 10 years. If a farm is utilizing the tractor close to capacity (it's not spending most of its time sitting in the barn), this purchase makes sense. A smaller farm that is just getting started might not be at a large enough scale of production to afford this expense. In these circumstances, it makes more sense to rent. While renting does involve some trade-offs, it allows you to only pay for the capacity that you need, when you need it.

In general, it is important to take a balanced approach to spending. Overspending reduces profitability and means more cash outflows, but underspending can also cause problems. Many spending categories are absolutely necessary and are at their core investments, meaning their cost will be offset directly by future resulting income. This is commonly called "return on investment," or ROI. A farmer who skimps on needed fertility inputs or labor to control weeds could be reducing yields and potential sales to a far greater degree than the money saved. The old adage "You must spend money to make money" is often true when investing in needed farm labor, inputs, supplies, equipment, and infrastructure.

Farmers Share Their Experience

FARM FINANCING: PERSPECTIVE ON GROWTH AND CAPITAL INVESTMENTS

WE OFTEN RAISE THE QUESTION of how much risk you are willing to take on in managing your farm business. After going through this chapter, you may have new perspectives or insights into financial risk and what financing options you are most comfortable with. Below you will learn how several farmers have worked through this issue of financing, from start-up to growing the business. How would you compare each farmer's tolerance for risk?

SARAHLEE LAWRENCE, RAINSHADOW ORGANICS

(profile on page 218)

In terms of building the farm and my infrastructure, I wanted to start out with a minimum amount of input because I wasn't sure if I was going to like farming.

MELANIE KUEGLER AND CHRIS JAGGER, BLUE FOX FARM

(profile on page 168)

Melanie: We've taken the approach of not going too far into debt at one time. At the end of each year, we've sat down and looked at our profit for that year, which has helped us determine how much we can put into infrastructure for the coming year.

Chris: We've always been a pay-as-you-go kind of farm just because we didn't want to get in over our heads. If we wake up one morning and realize we just can't keep doing this, we don't want to have a ton of debt hanging over our heads and go under. It's not that we feel that would happen; that's just how we've worked.

Paying as we go has been really good for us, too, because it has prevented us from growing our operation too quickly. We really have to incrementally increase our scale.

Our home property where we started as well as our other piece of property are owned by a family trust. We've put that into an LLC, which oversees just the pieces of land. If we didn't have Melanie's parents behind us in that way, I don't think we could have done it from the start. But at the same time, we have to work hard to make the mortgages on both properties.

ENTERPRISE ANALYSIS

A profit and loss statement is a great place to begin looking for the costs that might need attention. Eventually though, you will want to look at how much each individual enterprise on your farm contributes to (or detracts from) your bottom line. To do this, you must break the farm into separate enterprises. You can think of enterprises as income-generating centers within your business — usually specific products, market channels, or other areas of the farm, like a greenhouse that supplies seedlings for field production or a breeding program that supplies piglets. Looking at the income or cost reduction generated by a single enterprise on the farm and comparing that to the costs that are specific to that enterprise can give you key insights into how you might improve the profitability of your farm and whether or not certain products are worth producing.

Calculating the production costs for individual products can also identify the prices you need to charge to cover your costs. In general, some level of competition-based pricing is unavoidable in our market-based economy, but knowing your costs can help you make the argument to your customers that you need a higher price.

Calculating specific costs of production is beyond the scope of this book. Most farmers typically need a few seasons to get their systems established before they are ready to focus on the record keeping required to determine

PROFIT BENCHMARKING

We define profit as the difference between your income and your expenses. In the income and expenses budgeting section of this chapter, we refer to profit as "net income." Unfortunately, it is not quite as simple as it looks. Not everyone includes the same categories in income and expenses when calculating profit. This is especially true when it comes to including the cost of paying yourself, the farmer, for the time you put into your farm business. If you are a sole proprietor, technically any money you take out of the business for personal use (commonly called an owner's draw) is not an expense. To analyze profitability, you might want to track this differently, so that your labor is counted as an expense. On the other hand, businesses that are incorporated (see the section on Business Structures in chapter 6, "Keep It") are required to pay anyone, including the owners, as employees. In that case, even owner labor would be considered an expense and the profits would look very different, even if the farms were otherwise exactly the same.

Profit calculations are often made pretax, but not always. To get a realistic idea of how much money you will actually have at any point during the year, it's a good idea to include estimated tax payments in your budget.

One way to gauge how you are doing is to compare your profits (and other metrics) to other farmers' numbers, but you need to be careful when doing this. You must understand what other farmers are including and what they are leaving out. Another good option is to compare profits and other metrics in your own farm business from one year to another.

their own costs for specific enterprises. When you are ready, refer to the Resources section for a list of suggested tools and training materials that can help you with the costing process. In the meantime, enterprise budgets developed by land-grant universities can help you estimate the costs for individual enterprises. Several cost-study and enterprise budget websites are listed in the Resources section, or search the Web for "crop enterprise budgets" or "farm enterprise budgets" with your state's name.

Analyzing Cash Flow Statements

The main purpose of a cash flow budget is to reveal future cash shortages so that you can plan ahead and come up with strategies to avoid them. Cash flow statements tell you when things are going as planned and when they are not (which is most often the case). For this reason, it is important to monitor and analyze cash flow regularly (at least monthly) so that issues can be addressed in real time. (This is different from either the P&L or the balance sheet, which are analyzed at the end of a full production cycle — typically a growing season.)

The table below shows a simplified cash flow statement. In this example, the farm started with $10,000 in cash. Significant cash outflows (from expenses) occurred in January, and February and March had no cash inflows (from sales). In the first three months of the year, more cash went out than came in — meaning the farm's cash flow was negative. By June, this trend reversed and cash inflows (from income) started to outweigh cash outflows — meaning cash flow was positive. This trend continued for the rest of the year. This cash flow statement shows no months with a shortfall, so the farm does not need to consider any strategies to avoid cash shortages in the future. But what if it did? What if it was late spring, and cash started coming in later and more slowly than expected?

SAMPLE CASH FLOW STATEMENT

MONTH	STARTING CASH	CASH IN	CASH OUT	ENDING CASH
January	$10,000	$0	$1,000	$9,000
February	$9,000	$0	$2,000	$7,000
March	$7,000	$0	$3,000	$4,000
April	$4,000	$1,000	$3,000	$2,000
May	$2,000	$2,000	$3,000	$1,000
June	$1,000	$4,000	$3,000	$2,000
July	$2,000	$6,000	$3,000	$5,000
August	$5,000	$8,000	$4,000	$9,000
September	$9,000	$9,000	$4,000	$14,000
October	$14,000	$6,000	$2,000	$18,000
November	$18,000	$6,000	$1,000	$23,000
December	$23,000	$3,000	$1,000	$25,000

STRATEGIES TO IMPROVE CASH FLOW

Borrowing. Tight cash flow requires some farm businesses to utilize loans or short-term credit. This is especially true for more immediate cash flow issues when you can't employ other, longer-term strategies. It takes time to apply for and secure a loan or a line of credit. A loan will give a one-time lump sum, while a line of credit will give you less money over time but will give you more flexibility about how the funds are used and paid back. The borrower can draw down on the line of credit at any time, as long as he or she does not exceed the maximum set in the agreement. The advantage of a line of credit over a regular loan is that the borrower doesn't usually have to pay interest on the credit that is unused, and the borrower can draw on the credit as needed. Some credit terms require that the outstanding balance be paid immediately at the financial institution's request. Credit cards are lines of credit, but sometimes banks or credit unions offer better rates to businesses than major credit cards do.

Advanced payment arrangements. Sales arrangements where farmers charge customers in advance can help cash flow. Community-supported agriculture is one such approach (described in more detail in chapter 3, "Sell It"). Certain arrangements for direct marketed meat products may also require customers to pay in advance. Advanced payment arrangements may have higher administrative costs and/or a limited customer pool. This is a long-term strategy, since it takes time to develop these types of sales arrangements. Note that taking payments in advance obligates you to deliver in the future, so be clear with your customers about the risks in the case of a crop failure or lower-than-expected yields.

Terms on receivables. Take steps to get paid quickly by setting strict and short-duration payment terms on restaurant and wholesale accounts. Some small farms request that restaurants pay upon delivery. It may be more challenging to negotiate payment terms with large-scale sales outlets because these businesses are more likely to dictate their own terms. Each farm business must evaluate how strictly they want to enforce payment terms, and if they want to impose penalties such as late fees, interest, or dropping delinquent accounts. Accepting electronic payment by bank wire or credit card may in some instances improve cash flow, but it can also result in additional fees. Establishing or changing terms on receivables might take less time than developing new payment arrangements, but it is still a long-term strategy for addressing cash flow issues. In some wholesale markets, it is not uncommon for farmers to wait months to be paid for invoices.

Analyzing Balance Sheets

The statement of financial position, otherwise known as the balance sheet, is less useful as an internal management tool than the profit and loss statement and the cash flow statement. However, it does show how net worth (owner's equity) changes from the beginning of one period to the next.

The sample balance sheets opposite show two simplified balance sheets, one from the beginning of the year and one from the end. Notice that both assets and owner's equity have increased by $10,000 from the beginning of the year to the end. This increase is a result of profits that have been retained in the business.

When you are just starting out, it may feel like there is little financial reward for all your hard work. But remember that investments in equipment and infrastructure are actually increasing your owner equity from year to year. Farmers in this situation should look at their balance sheet periodically to get a true picture of their financial position. They may be doing much better than the profit and loss statement indicates.

SAMPLE BALANCE SHEETS

	BEGINNING OF YEAR	END OF YEAR
Assets	$60,000	$70,000
Owner's equity	$40,000	$50,000
Liabilities	$20,000	$20,000
TOTAL ASSETS	$60,000	$70,000
TOTAL EQUITY + LIABILITIES	$60,000	$70,000

FINANCIAL RATIOS

Financial professionals use a number of ratios to assess the financial health of your business to determine if they will lend to you. Each ratio looks at a slightly different aspect of the financial picture, so you need to calculate and interpret them very carefully. The actual numbers in a ratio can give you some indication of a farm's financial situation, but more commonly, ratios are evaluated against certain benchmarks that indicate the financial health of the business. These benchmarks vary according to the type of business you are running and the stage of the business's growth. Some financial ratios are used in relation to each other, and that relationship is more important than the absolute numbers. Tracking and comparing the ratios over time can give you an indication of how your business is improving or where problems might be developing.

The current ratio, the ratio of current assets to current liabilities, is one financial ratio that lenders use. In this case, "current" means short term — generally less than one year.

Obviously, you want your current assets to be higher than your current liabilities; otherwise, you cannot pay your bills and are in imminent danger of failing as a business. General financial guidelines might say that a ratio of 1.5 or better is good, and if you are between 1.0 and 1.5 you are okay, but under 1.0 signals trouble.

When calculating and analyzing any financial ratio, be consistent and compare it to numbers that are calculated in the same way. For example, with the current ratio, are you consistent in how you identify what is "current" and what is longer term? Did you just purchase a piece of equipment that is making your current assets look lower or your current liabilities look higher than usual?

If you are interested in looking at more ratios and you have a good accounting system, it should be relatively easy to pull out the numbers that you need to calculate them, but make sure you understand clearly how the parts of the ratios are defined and what the ratios are indicating before trying to use them. The book *Fearless Farm Finances* from MOSES (Midwest Organic & Sustainable Education Service) has a detailed section on ratios for farmers that will get you started.

WORKSHEETS FOR BUSINESS MANAGEMENT

With a better understanding of business management practices for your farm business, this is a good opportunity to review several of the farm plan worksheets you have worked on in previous chapters. In particular, take a look at the Values and Vision (appendix 1-5) and Ideal Farm (appendix 1-4) worksheets from chapter 1, "Dream It." In addition, we provide five farm plan worksheets specifically related to what we have covered in this chapter (appendices 1-19 through 1-23).

BUDGETS AND VARIANCES

The Budgets and Variances worksheet (appendix 1-19) will help you prepare a whole farm budget for your business. Remember to clearly define the parameters and goals of your budget (such as type of farm, product, acreage, start-up, seasonal, and expansion). If you are in the very beginning stages and are not yet operating a farm business, create a hypothetical budget based on your current plans and projections. Remember, you don't have to complete this in one sitting. Do what you can, then return to it later as you have more information.

CASH FLOW AND ACCOUNTING SYSTEMS

The Cash Flow and Accounting Systems worksheet (appendix 1-20) gives you the opportunity to evaluate your cash flow situation and also describe the accounting system that you have set up or that you think will work best for your farm.

PRODUCTION COSTS

The Production Costs worksheet (appendix 1-21) will help you assess how well you are managing costs for your farm business. If you cannot complete the table at this time, that's okay. Do what you can, then return to it later as you work through the remainder of the chapters.

FINANCING

The Financing worksheet (appendix 1-22) is designed to assess current and future financing options. If you are farming now, indicate if you currently use a source of financing, how much debt you currently hold in each financing category, and when you hope to pay off that debt. Also indicate if you want to pursue that source of financing in the future. If you are not yet farming, indicate if you want to pursue that source of financing and provide an estimate for how much financing you will be seeking.

BALANCE SHEETS

The Balance Sheets worksheet (appendix 1-23) will help you create a balance sheet for your farm business. If you are already farming, use current figures for your farm. If you are not yet farming, try developing a hypothetical balance sheet using the best data you have.

It's Not All about Dollars

When analyzing the financial situation for your farm, remember that dollars are only one measure of success. While it is important to make a profit (or at least break even), it is also important to consider quality-of-life issues for yourself and your farm team and to consider the health and sustainability of your community and the natural environment where you live and work. These considerations may not have dollar numbers attached to them, but they are important for the long-term sustainability of your farm business and should be factored into your decision-making.

PUTTING IT ALL TOGETHER

Many farm businesses are successful over the long term and, in time, pass from one generation to the next. What is the secret to their success? There is no one answer to that question, and even businesses that have been successful in the past may find themselves struggling as circumstances change. Nonetheless, farmer experience points to several management tips that can help keep you moving in the right direction.

BUSINESS MANAGEMENT TIPS FROM SUCCESSFUL FARMERS

- Be cautious about borrowing, but do not undercapitalize the business. Many businesses fail because they are undercapitalized.
- Expect frequent challenges in the early years of the business, with steep learning curves and high start-up costs.
- Always overestimate expenses and underestimate income to account for unforeseen challenges like equipment breakdowns, effects of bad weather on crop yields, or weak markets.
- Do not overspend on materials, supplies, and equipment the farm does not really need.
- Understand that direct marketing strategies require time and resources, thereby affecting your overall costs.
- Be cautious of early-season cash flow issues.
- Establish reasonable ways to pay yourself for your work.
- Emphasize the production of the most profitable crops; know which ones they are.

BLUE FOX FARM
ADAPTING TO CHANGE

OREGON

Blue Fox Farm

Blue Fox Farm is a 67-acre diversified vegetable farm in the Applegate Valley of southern Oregon. Melanie Kuegler and Chris Jagger have been farming there since 2003.

Chris and Melanie did not have any exposure to farming while they were growing up. They were town kids. Chris went to college at Iowa State University. "I started getting curious about where my food was coming from and started tracing it back until I got to production. And then when I got to production, I was really intrigued by that whole side of it and feeding people." Melanie says she grew up the first 18 years of her life barely knowing what a farm was. "I went to college in Santa Cruz, UCSC, and started doing some work at the farm on the campus and doing some internships at smaller farms around the Santa Cruz area. And that's where I fell in love with it."

Melanie and Chris started out as farm managers in California and Colorado, then they began looking for their own farm in 2002. They were attracted by the beauty of southern Oregon and did their homework to research the business potential in that part of the state. "When we found this piece of property, we just went for it. We had just enough money to put a down payment on the land and one tractor. We started with an acre and a small booth at two farmers' markets," says Melanie.

"The second year we farmed about two acres," says Chris. "It was one of those grueling years, but it was good that we went through it to get a true visual imagery of what the whole system involved. So we're fortunate to have been able to start from the ground up." From the initial production on their home property, Melanie and Chris expanded on a

mix of purchased and leased land until they had 35 acres in production. After farming at that level for several years, they have now scaled back to producing on about 15 acres each season.

They sell their produce through a variety of market channels including farmers' markets, wholesale to grocery stores, and direct to restaurants. They also operate what they call a Market Bucks program where community members invest in the farm at the beginning of the season and, in return, get credits that they can use to buy produce at the Blue Fox farmers' market stand at a reduced price. They also sell to a wholesale distributor (Organically Grown Company) out of Eugene, Oregon.

"Our vision has changed because we've grown and changed," notes Melanie. "We've started a family, which means I am at home with the two children a lot and we have to hire other people to help. I can't be out there working the hours that I was working before. And I think things have changed because we're responding to what the markets want, so what we're growing and how much we're growing has definitely evolved over the years."

Melanie and Chris attribute much of their success as business owners and managers to their adaptability. "Part of our discussion in our down period each year in December is, How do we look at the next year? What has changed? And how do we adapt to it? And we've seen a lot of success so far, year to year, I feel, primarily because of that," says Chris.

"We want people to get good-quality fresh produce. And we want to do that while being able to afford what we need for our family in terms of good food, good clothing, good housing, health care, et cetera. And we want that for our employees. So our highest value is making sure that we're all taken care of while producing good product for people," says Melanie.

GROW IT

GOODFOOT FARM

CHAPTER FIVE

MANAGING THE WHOLE FARM ECOSYSTEM

Let's review some of the key elements needed to create a successful farm business.

- A clear vision and mission (chapter 1, "Dream It")
- Self-awareness and respect for the skills and talents of your farm team (chapter 1, "Dream It")
- Thorough knowledge of your farm's natural resources (chapter 1, "Dream It")
- Good planning, with a whole farm perspective (chapter 1, "Dream It")
- The ability to make informed decisions about equipment and infrastructure (chapter 2, "Do It")
- Skills to hire, motivate, and manage employees (chapter 2, "Do It")
- Knowledge of your marketplace and a clear understanding of what your customers are looking for (chapter 3, "Sell It")
- A marketing plan that helps you decide what marketing channels work best for your situation (chapter 3, "Sell It")
- Flexibility and the ability to adapt to rapidly changing market conditions (chapter 3, "Sell It")
- Basic business management skills that enable you to project income and expenses before the start of the growing season and plan for sufficient cash to cover your monthly expenses (chapter 4, "Manage It")
- An accounting system and accurate record keeping to analyze your financial situation (chapter 4, "Manage It")

In this chapter, we add the following obvious, though nonetheless critical, item to the list: the practical foundation for growing crops and raising livestock. Given the diversity of farming and ranching systems, and the regional and climate variations across the United States, the goal of this chapter is to introduce the basic principles and practices of sustainable agriculture.

As a farmer, you are the manager of a particular ecosystem — *your* farm ecosystem. That ecosystem is complex. It has many different components (soil, water, air, organisms, biogeochemical processes) that interconnect in a variety of ways and are finely tuned to environmental conditions, and that connect to all other aspects of the whole farm. This complexity presents you — the farm ecosystem manager — with both challenges and opportunities.

Eliot Coleman takes this concept a step further in his book *The New Organic Grower: A Master's Manual of Tools and Techniques for the Home and Market Gardener*. In that book, he describes the complexity of a farm, and the farmer's role, using the analogy of an orchestra conductor.

> *Working with living creatures, both plant and animal, is what makes agriculture different from any other production enterprise. Even though a product is produced, in farming the process is anything but industrial. It is biological.*
>
> *The skills required to manage a biological system are similar to those of the conductor of an orchestra. The musicians are all very good at what they do individually. The role of the conductor is not to play each instrument, but rather to nurture the union of the disparate parts. The conductor coordinates each musician's effort with those of all the others and combines them in a harmonious whole.*
>
> *The major workers — the soil microorganisms, the fungi, the mineral particles, the sun, the air, the water — are all parts of a system, and it is not just the employment of any one of them, but the coordination of the whole that achieves success.*

To help you develop your skills as a farm ecosystem manager — a "conductor," using

Coleman's analogy — we address three major topics in this chapter:

The farm ecosystem. To manage a farm ecosystem, you first need to know the basic components of that system and how they work together to form a whole. In this section, we explore the different cycles that comprise all ecosystems and the ways the cycles interconnect.

Managing the whole farm ecosystem. Farm ecosystems *can* be managed to improve the productivity and profitability of the farm. But how do you do that most effectively? This section uses the concept of "working with nature" to present basic guidelines that will help you become an effective farm ecosystem manager.

Key practices in sustainable agriculture. This final section covers how to put ecosystem knowledge into practice. You will learn about six key practices that are common in many sustainable farming systems: cover crops, crop rotation, conservation tillage, compost and soil amendments, rotational grazing, and farmscaping. As you explore these practices, you will see how they relate to the various ecosystem cycles you learned about in the early sections of the chapter, and how they contribute to the long-term health of your farm or ranch system.

BY WORKING THROUGH EACH SECTION, YOU WILL:

1. View your farm as an ecosystem and understand your role as a farm ecosystem manager
2. Learn about the natural cycles and organisms that are part of your farm ecosystem
3. Understand the basic principles of managing a farm ecosystem and apply them to your own farm or ranch to enhance its productivity and profitability
4. Become familiar with several of the key practices used in sustainable farming systems

PLANNING AND INTERVENTION

We begin by offering a practical framework for thinking about farm ecosystem management. Consider that many of the tasks you do as a farmer fall into two categories: planning or intervention.

Planning. Planning encompasses both long-term and short-term (seasonal) production strategies for the farm. For example, in developing a whole farm plan, you think through your vision, mission, and goals; you assess the land and other natural resources that comprise the farm ecosystem; and you work on the business side, evaluating your marketing strategy and developing skills to manage the business wisely. This planning is important, and the key to success is maintaining that perspective over the long haul and working continuously to improve the farm system. This same planning perspective applies to specific crop or livestock management practices. The goal is to create a healthy farm ecosystem that remains productive in the face of pest pressures, weather extremes, changing climate, and other challenges. To develop that kind of system takes planning. In this context, planning can be seen as the key step in preventing or avoiding problems before they occur, which in turn leads to farm resilience.

Intervention. Problems arise in even the best-designed farming systems: weather extremes, pest problems, nutrient deficiencies, or other soil-related problems. Intervention is the work farmers do during the growing season to address and correct these problems. We usually think of these as targeted interventions (such as cultivation to remove weeds, applying fertilizer to correct a nutrient deficiency, and increased irrigation during hot weather), but they could also include adjustments to your long-term plan to avoid or prevent that problem in the future (for example, installing row covers to protect crops from frost, or managing soil fertility to avoid deficiencies).

Deciding when and how to intervene requires an accurate description of the problem. Getting that clear picture may involve activities such as monitoring soil conditions and crop health, monitoring fields and borders for pests, accurately identifying pests, and using weather forecasts and degree-day models to predict potential problems in advance. For livestock producers, you may need to conduct regular animal inspections to identify and monitor emerging problems. Proper identification and diagnosis are crucial: responding to the wrong problem wastes resources and can cause additional problems while the original misdiagnosed problem continues unchecked. Accurate record keeping goes hand in hand with these tasks.

Pest management provides a good example of this interplay between planning and intervention. In sustainable farming systems, the four basic components of an ecologically based pest management strategy can be summarized within the planning-intervention framework as follows, adapted from *Building Soils for Better Crops* (see Source Citations):

PLANNING PRACTICES

1. Manage the farm to prevent or avoid pest problems: create conditions for healthy crops/livestock that are able to resist or outcompete pests.
2. Make it difficult for pest organisms to grow and reproduce (stress the pests).
3. Boost populations of beneficial organisms that keep pest organisms under control.

INTERVENTION PRACTICES

4. When pest problems arise and intervention is required, choose control methods that are environmentally friendly with minimal disruption to the farm's ecological balance. This requires accurate identification of pests, monitoring, and, if necessary, proper selection and timing of pest-control materials.

Top: Solar insect trap at Kiyokawa Family Orchards to monitor pest pressure
Bottom: The night's catch for analysis

Taking this concept a step further, consider what types of practices might be included in each category:

PLANNING PRACTICES

- **Crop rotation.** Well-planned rotations can improve soil quality and reduce pests (i.e. insect, disease, weeds) by interrupting their life cycles.
- **Cover crops.** Including cover crops in your rotation can improve soil structure and fertility; cover crops can also enhance populations of beneficial insects.
- **Cultivar selection.** Select crop varieties that are adapted to your location and soil and that are resistant to pests.
- **Plant spacing, arrangement, and pruning.** Plant spacing and architecture can alter the microenvironment within the field, which can prevent certain pest problems.
- **Planting and harvesting dates.** Timing of planting and harvesting can affect the ability of crops to avoid pests and withstand extreme weather. Planting earlier or later can avoid pests that are active at certain times.
- **Cultivation.** When properly timed, tillage is an effective practice for preventing certain weed problems; incorporating crop residues into the soil can keep certain pest problems in check.
- **Wildlife habitat.** Create habitat by farmscaping for beneficial insects and other wildlife that can reduce pest problems.
- **Proper irrigation management.** Good irrigation scheduling can reduce pest problems, improve nutrient availability, and enhance crop performance.
- **Soil management plan.** Develop an overall soil management plan to enhance soil health and quality. Healthy soil makes your farm and your crops more resilient.

INTERVENTION PRACTICES

- **Monitoring.** Monitor weather conditions, crop and livestock health, pest populations, and soil nutrients so that you notice potential problems before they become serious and can intervene before yield and quality are badly affected.
- **Appropriate use of fertilizers and other soil amendments.** This promotes plant health and reduces some pest populations while protecting water quality.
- **Cultivation.** Sometimes, emergency cultivation may be needed to reduce an unexpected weed infestation or to prevent an aggressive weed from going to seed.
- **Row covers.** Use row covers to protect young plants from pests or frost damage.
- **Biological pest controls.** Use mating disruption techniques for insect pests and enhance natural biological controls.
- **Right pesticides.** Use environmentally friendly, low-toxicity pesticides that comply with marketing standards, such as organic certification.
- **Correct plans.** Adjust your plans for next year based on lessons learned this year.

Keep in mind that there is overlap between planning and intervention. Some practices appear in both categories, and as noted above, an intervention might (and often should) include an adjustment to your long-term plan or strategy. This is part of the complexity of farming. Keep these two tasks — planning and intervention — in mind as we begin our exploration of the farm ecosystem.

Farmers Share Their Experience
PERENNIAL PRODUCTION STRATEGIES

YOU WILL GAIN a better understanding of how planning and intervention play out within the context of the farm after reading these comments from Randy Kiyokawa. Note how Randy has planned for the long term and what factors have figured into his decisions. He also describes the kind of information and data that inform his day-to-day management choices, and how he constantly reevaluates and fine-tunes his system.

RANDY KIYOKAWA, KIYOKAWA FAMILY ORCHARDS (profile on page 94)

Parkdale is unique. It has a great climate for growing apples and pears. I feel that we have some advantages in that we don't have the heat units that cause a lot of insect populations to go through many generations.

There are some drawbacks in that we do have a short growing season, so I struggle with certain varieties of apples, like Granny Smith and Braeburn. But because of our soils and the cool nights, we grow some of the best Honeycrisp, and other varieties do really well here.

One thing that's different with orchards is that you plant an apple tree and you don't expect anything from it for 2, 3, or 4 years. For pear trees, it's 7 to 11 years. That's very different from planting an annual crop and getting some revenue right away. So we're planning, I'm thinking, three, four years down the road at least. Overall, I look at the long term.

How do I plan for production? I don't want to have my crops come on all at once. That's why it's nice to have the peaches or the cherries earlier in the season, before the rest of the fruits. I have planted some here and some three miles from here, and just because of the elevation changes, their harvest dates are about three days apart.

I'm really booked. I have a hard time picking all the fruit and berries when they're ready, so if I'm going to plant another variety or crop, it'll be to extend my operations more into summer. I don't need to extend it into winter because winter comes quick enough here in Parkdale.

We do a lot of soil sampling, at least once a year (we sample weaker blocks more often), because our soils are very depleted in phosphorus. We also do tissue sampling if needed. A lot of my production strategy is about knowing the health of the tree. Since I have been looking at the trees on the farm for a number of years, I know when something's not right, and that helps. From there, I go to the experts and get the advice I need.

We do a lot with soil fertility. My father and grandfather put on a lot of nitrogen. I think growers this day and age are spoon-feeding their crops just what they need at the time they need it, so they don't leach out nitrogen or starve a plant of something that the plant would need to produce a good fruit.

Disease and insect management is one thing that I don't have enough time to monitor so I have a field man come out and scout my orchard. And he checks for diseases and insects. He uses degree-day models that are set up through weather stations. He gives me recommendations.

THE FARM ECOSYSTEM

An ecosystem consists of all the organisms — plants, animals, fungi, and various types of microorganisms — living within a specific area, interacting with each other and their environment. An ecosystem can be characterized by the diversity and abundance of organisms within its boundaries, where they live, and how they interact with and depend on one another. That complete picture of how organisms fit into an ecosystem is called the food chain or food web — terms that you are probably familiar with. Microorganisms and the microbiomes in the soil and on plants are especially important components of the food web. Supporting and fueling those interconnections are the various natural cycles of water, energy, carbon, and many other chemical compounds (including plant nutrients) that occur within the environment. The ecosystem concepts and biogeochemical cycles introduced in this section are adapted from the USDA SARE course Agricultural Ecosystem Management (see Source Citations).

Farms and ranches are a type of ecosystem. They receive the same natural inputs as any other ecosystem — energy and heat from the sun, carbon, oxygen, water, and nutrients. But farms and ranches differ from natural ecosystems in that they are *managed* ecosystems, with the goal of producing a product that is removed from the farm for sale and consumption. In this way, energy and nutrients are removed from the system (much more so than in natural ecosystems). Farmers generally add other inputs to the system to compensate for this loss (such as added nutrients from fertilizers, cover crops, compost, or other amendments).

Based on this concept, what approach to managing the farm/ranch ecosystem is best? In sustainable agriculture, the goal is to work with nature as much as possible. This requires some awareness and knowledge of the biogeochemical processes that are at work in the farm ecosystem, including energy flow, the carbon cycle, the water cycle, and nutrient cycles (such as the nitrogen cycle).

In addition, farmers must be concerned with the life cycles of all the organisms that live within the farm ecosystem: crops and livestock, weeds, arthropods, and microorganisms (particularly plant and animal diseases). Let's explore these biogeochemical processes and seasonal life cycles in more detail.

Biogeochemical Cycles

One of the key characteristics of the earth's ecosystem (biosphere) is that the chemical compounds that comprise the system are conserved and recycled. The transformation pathway that a compound follows is called a geochemical cycle. In a moment, we will address three of the most important biogeochemical cycles in agriculture — carbon, nitrogen, and water. First though, we look at the energy flow that powers these cycles.

ENERGY FLOW

All organisms need a regular supply of energy to grow and reproduce. In an ecosystem, organisms are classified as producers or consumers based on how they get their energy. Plants are *primary producers* since they are able to take energy from sunlight and convert it into carbohydrates and proteins that are stored in the form of biomass — leaves, shoots, roots, fruits, and seeds. *Consumers*, on the other hand, get their energy from consuming other organisms (plants and/or animals). One way of categorizing consumers is based on what they eat: herbivores, carnivores, and omnivores. Another important group of consumers comprises *decomposers*: organisms that live in the soil and consume dead plant and animal material. As energy flows through an ecosystem, it is used by organisms for various life processes or it is lost as metabolic heat. This flow of energy through all the organisms in an ecosystem is related to the concept of the food chain or food web.

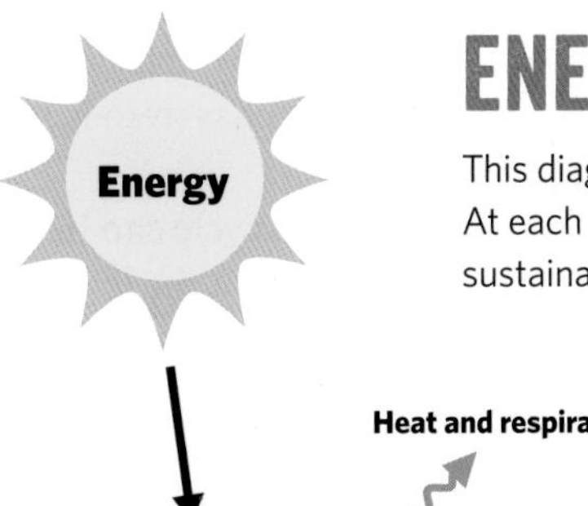

ENERGY FLOW ON A FARM

This diagram shows how energy from the sun moves through a farm ecosystem. At each level, some energy is lost through heat and respiration. A goal of sustainable agriculture is to capture and use as much of this energy as possible.

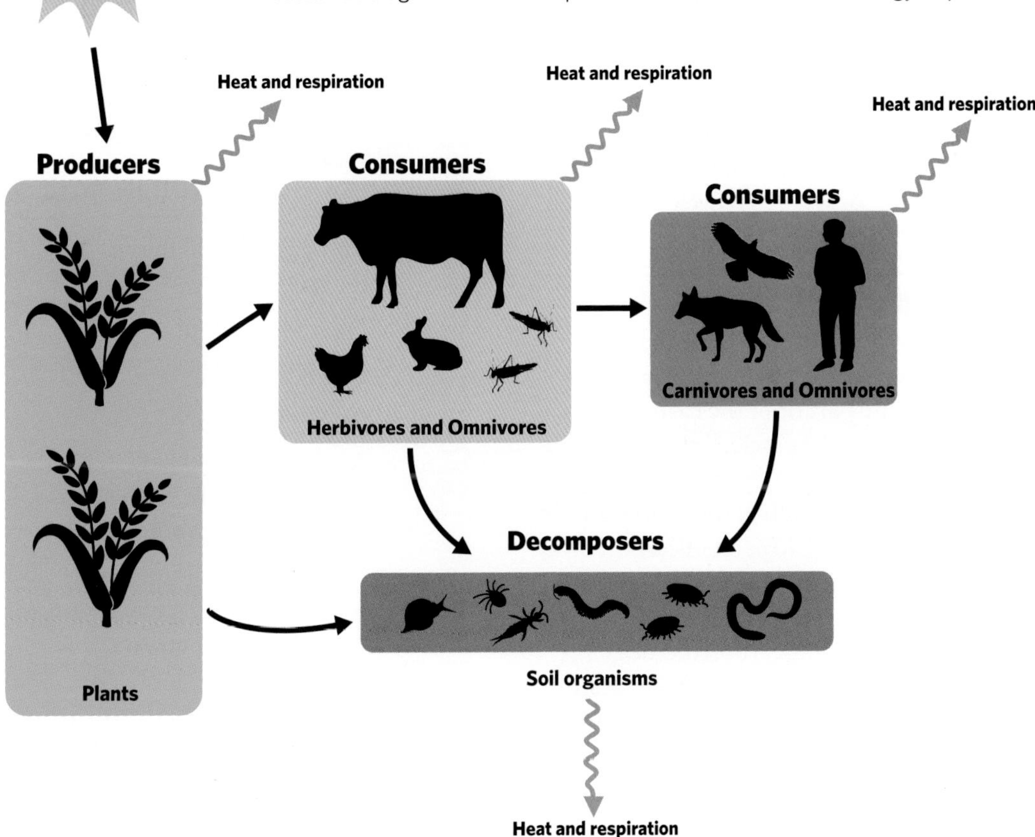

The loss of mass (carbon) and energy within a farm ecosystem and food web is magnified because the plants being grown (the crops) are removed from the farm ecosystem instead of being cycled back to the soil, as most of the plant material would be in a natural ecosystem. Compounding this inefficiency in the farm ecosystem is the need for additional energy inputs such as fuel for equipment and machinery, nutrients, and electrical energy to grow, harvest, process, and market the crop. An important task for sustainable farmers and ranchers is to manage energy and nutrient cycles to be as efficient as possible.

Think about your own farm or ranch ecosystem and how your understanding of energy flow translate into specific management goals. One goal might be to capture as much solar energy as possible. For example, rather than leaving fields fallow, you might include cover crops in your rotation to produce vegetation (organic matter) during the off-season that can be returned to the soil, or use optimum crop planting densities that increase the amount of soil cover. Another goal might be to reduce energy loss, or to simply improve the overall energy efficiency of the farm ecosystem. If you have livestock, that might include looking for ways to recycle animal wastes back into the farm system.

THE CARBON CYCLE

The carbon cycle is closely related to the concept of energy flow. All living organisms rely on — and are part of — the carbon cycle. Carbon is exchanged or cycled among oceans, the atmosphere, and ecosystems. Carbon forms the backbone of all organic compounds.

Plants are generally 40 percent carbon on a dry weight basis. Plants capture carbon from the atmosphere during photosynthesis. Plants take in carbon as carbon dioxide (CO_2) through small openings (stomata) in their leaves. They use solar energy through photosynthesis to make carbon-based molecules such as cellulose, protein, sugars, starches, and lignin. Plants "burn" some of these compounds as energy sources (a process called respiration) for growth and reproduction. As plants shed leaves and die, their carbon is added to the soil, where it contributes to soil organic matter. All life on earth is dependent on these processes of photosynthesis and respiration.

Animals consume living and dead plant material and "burn it" as a food source. The carbon combines with oxygen as it is used as an energy source, becoming carbon dioxide, which in turn is released into the atmosphere as these organisms breathe. The carbon dioxide is again available for plants to take in and repeat the cycle. Fungi and many types of microorganisms play a key role in this process. The carbon cycle in a farm ecosystem is illustrated at right.

Knowledge of the carbon cycle can help you utilize methods to improve the level of

WAYS TO INCREASE ORGANIC MATTER IN SOIL

- Return crop residues to the soil.
- Plant cover crops.
- Reduce tillage.
- Apply compost.
- Apply raw organic wastes (such as manure, leaves, and packhouse or processing wastes).
- Keep the soil covered to minimize the risk of erosion.

THE CARBON CYCLE ON A FARM

Through photosynthesis, crops and other plants fix atmospheric carbon as plant structures and sources of energy. Crops, other plant residue, and manure decompose into organic matter, which releases carbon into the atmosphere to continue the cycle. The key benefit of the carbon cycle on farms is organic matter, which enhances soil tilth, aggregation, and fertility.

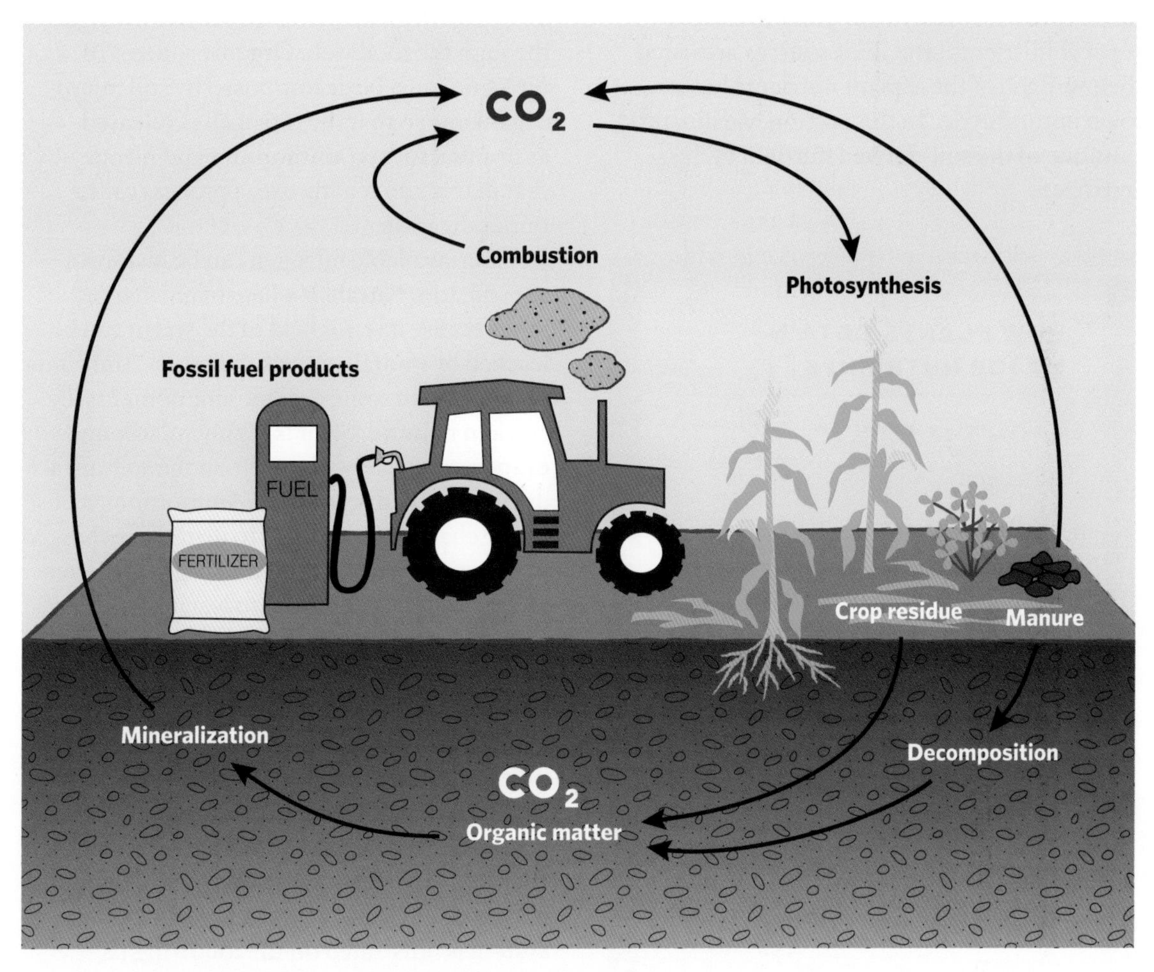

Adapted from Agricultural Ecosystem Management, Sustainable Agriculture Research and Education (SARE), USDA National Institute of Food and Agriculture (NIFA).

organic matter in soil. The list on page 180 is a preview of some of the steps we will look at in more detail in the Key Practices in Sustainable Agriculture section (page 203).

NUTRIENT CYCLES

Plants need certain nutrients in order to grow and reproduce. The major plant nutrients (macronutrients) and their sources are listed below. Each of these plant nutrients has its own natural cycle. In this section, we illustrate just one of the soil-derived nutrient cycles: nitrogen.

HOW PLANTS OBTAIN MAJOR NUTRIENTS

FROM THE SOIL

Nitrogen

Phosphorus

Potassium

Calcium

Magnesium

Sulfur

FROM WATER

Hydrogen

FROM THE ATMOSPHERE

Carbon

Oxygen

Nitrogen cycle. The atmosphere contains about 78 percent nitrogen gas (N_2). However, atmospheric nitrogen is not available to plants. It must first be converted to ammonia (NH_3), ammonium (NH_4), and nitrate (NO_3) by nitrogen-fixing bacteria. Rhizobia, living symbiotically in legume root nodules, are one example, but other types of symbiotic and free-living bacteria also convert nitrogen into these usable forms. Nitrogen can enter the cycle from other sources besides the atmosphere. Manure, decaying plant materials, and other organic materials recycle nitrogen through the food web. Organic sources of nitrogen must be decomposed by soil microorganisms so that the nitrogen is released as mineral forms (ammonium and nitrate) that plants can readily use, a process called mineralization.

Plant-available nitrogen can be lost from the soil, too. Nitrate is a negatively charged ion; because it is not held to the soil, it can be leached by rainfall or overirrigation. Ammonia is common in some organic amendments like chicken manure, but it is highly volatile and evaporates quickly when left on the soil surface during warm, dry weather. Ammonium, a positively charged ion, is held to negatively charged soil particles such as clays and organic matter and therefore is less likely to leach.

Under normal conditions, ammonia and ammonium are quickly converted to nitrate. But when soil is wet and oxygen is in short supply (anaerobic conditions), denitrifying bacteria reduce nitrate to forms that are not available to plants and that are lost to the atmosphere. This can be a source of greenhouse gases and contribute to air pollution and acid rain. The nitrogen cycle in a farm ecosystem is illustrated on the following page.

Nitrogen is one of the major plant nutrients and, along with other nutrients, is a primary building block for proteins that are required for plant and animal growth. Nitrogen is also a key element in chlorophyll (for photosynthesis). Working with the nitrogen cycle is another important facet of managing your farm ecosystem. Soil nutrient testing can show you how fertile your soil is. You can use this

THE NITROGEN CYCLE ON A FARM

Nitrogen is one of the major plant nutrients and is a primary building block for proteins that are required for plant and animal growth. Working with the nitrogen cycle is an important facet of managing a farm ecosystem and can make the farm more resilient and lower production costs.

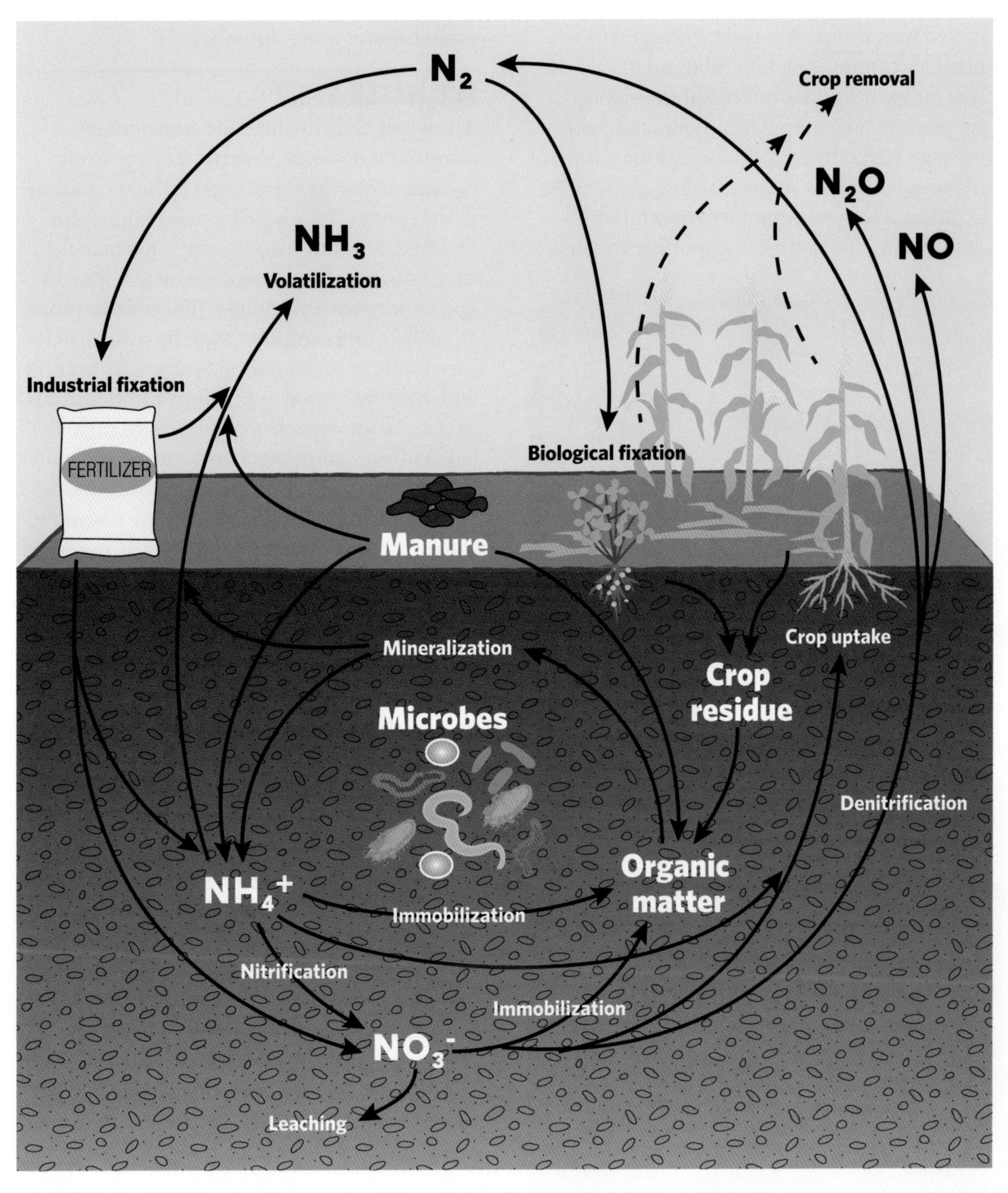

Adapted from Agricultural Ecosystem Management and *Nitrogen Basics — The Nitrogen Cycle* (see Source Citations).

information to provide the nutrients your crops need. If you have confirmed there is a deficiency, then you can decide how you are going to supply those nutrients.

Interestingly, the nitrogen cycle is closely linked with the carbon cycle. Nitrogen in plant and animal residues is bound to carbon molecules and is not readily available for plant uptake until it is mineralized. Soil microbes decompose organic residues. If there is enough nitrogen in the material, decomposition releases ammonia to the soil (mineralization); if there is not enough nitrogen, microbes scavenge ammonia or nitrate from the soil (immobilization). This principle becomes very important when you are dealing with organic materials added to the soil, such as crop residues, cover crops, compost, and other organic amendments.

THE WATER CYCLE

The water cycle involves the continuous movement of water on earth. Energy from the sun drives the water cycle by heating water in the oceans. This heating causes the water to evaporate, and rising air currents take the vapor up into the atmosphere, where it cools and condenses into clouds. Air currents move clouds around the globe, and the condensed vapor falls as precipitation. Some falls as snow, which can be stored annually or for thousands of years as snowpack, glaciers, or ice caps. In tropical and temperate zones, the precipitation flows over the ground as surface runoff, becoming lakes and rivers that return water to the oceans. Some water infiltrates through the soil into aquifers, which store huge amounts of water. Some stays near the surface, where it is absorbed by plant roots and transpired from leaves, adding water vapor to the atmosphere. Water can more easily enter soil with good tilth and structure. Soil that is also rich in organic matter has more water-holding capacity than depleted soil. Explore the water cycle in a farm ecosystem in the illustration on the following page.

Farmers and ranchers are always working with the water cycle. Since we are working with a natural system, it should be no surprise that the water cycle on your farm is connected with the carbon and nitrogen cycles. There are ways to manage carbon and nitrogen to improve the efficiency of the water cycle on your farm. Covering the soil with mulch can reduce soil moisture loss through evapotranspiration and soil compaction from rainfall.

THE WATER CYCLE ON A FARM

Farmers and ranchers can work with the water cycle to capture water when and where it is needed, and prevent runoff and leaching of nutrients into groundwater.

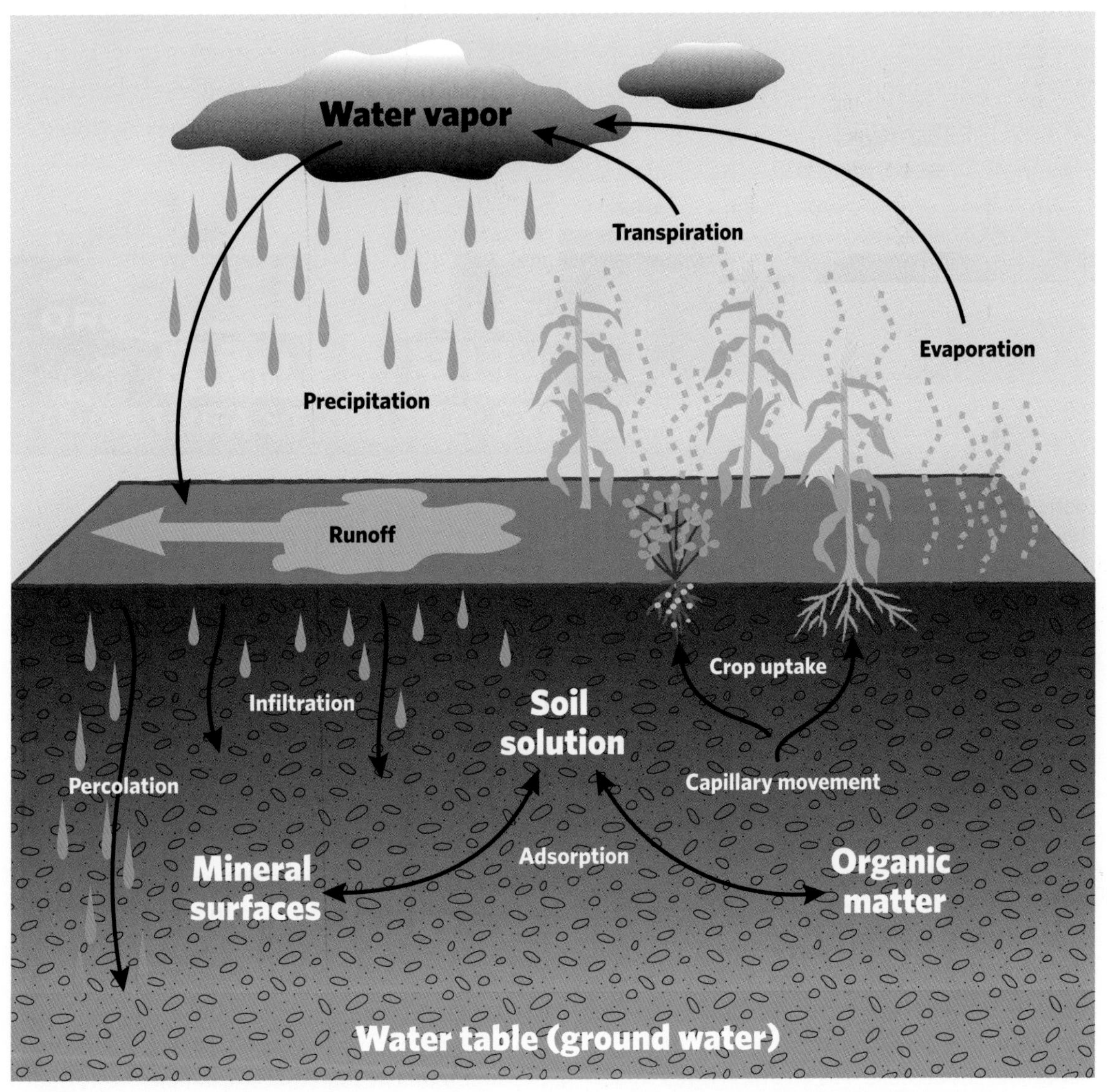

Courtesy of Sustainable Agriculture Research and Education (SARE), USDA National Institute of Food and Agriculture (NIFA)

Farmers Share Their Experience

INTEGRATED PRODUCTION STRATEGIES

THE ENERGY, WATER, CARBON, AND NUTRIENT CYCLES described in this chapter form the engine driving the production potential of your farm. Beth Hoinacki describes how this concept of production potential guides her decisions about her own farm. From her comments, note how Beth views the farm as a whole ecosystem, and how she includes herself in that perspective — her own needs as an individual and as a business manager.

BETH HOINACKI, GOODFOOT FARM (profile on page 44)

My job as a farmer is to enable my farm to express its potential. To do that, I must balance my needs as a farmer with the production potential of this piece of land.

I have Class II–type soils. Class II has some challenges, and we have to integrate those into our farming practices. We're on a north-facing slope, which is less than ideal in terms of capturing thermal heat. We're also in a cold microclimate. We're up against the Coast Range on the south side of the property here, so the cold air drains out of those hills and then it sits right down by the river.

We are consistently 10 degrees colder in the summer than down in the Willamette Valley, so we've had to come to terms with that and at the same time produce the things that we want to produce. I want to grow tomatoes, for example, but I don't like putting black plastic on the ground to grow crops, so the only way we can grow tomatoes here is in a hoop house. We'll never grow melons because it doesn't make economic sense to grow them in a hoop house. We'll never grow sweet corn, but I'm fine with that. Perennial crops do really well here, so they have been the foundation of our system. And then we put up hoop houses for season extension and for warm-weather crops.

The farm is as regenerative as possible given our potential resources. For example, we've integrated a lot of animals into the farm for fertility resources, pest and disease control, and as lawnmowers so that we're not having to buy fuel to mow the grass.

The biodynamic model asks you to realize the potential of your farm, to recognize that there are

Geese in postharvest blueberries eat any fruit on the ground, reducing a potential source of plant pathogens.

certain qualities that your farm possesses, and that in order for the farm to be productive you must keep things in balance. For us, that means not farming the slopes, keeping the slopes in permanent pasture or perennial crops, and farming just the level areas, which influences what we can produce on the farm.

One of the biggest challenges in perennial cropping systems is weed management. Weeds in the blueberries are a problem. You start weeding and soon you're saying to yourself, 'This is ridiculous. There's no way I can get through the field. I would just be weeding full-time, all the time, and I still wouldn't have a handle on it. There's got to be a better way.'

So we added chickens — they're basically little cultivators. One of the tenets of disease management in fruiting crops is that you clean up your field after the season because any leftover fruit can harbor pathogens. So we run chickens through the blueberries because they'll clean up all of our leftover fruit. They're my little sanitation workers. Then they scratch around in the berries and help cultivate and do some weeding. And then we throw some geese in there, too, because the geese are weeders. And then everybody will be fertilizing at the same time.

At the end of blueberry season in late August, the chickens and the geese go in. At that time of the year, the sheep are typically in the orchard keeping it relatively well mowed. We rotate the chickens and the geese every two weeks through the blueberries, so they work on a couple of rows at a time.

After the leaves fall, the sheep move into the blueberry field. They keep it mowed and munch on the blackberries that are volunteer weeds out in the blueberry field. Everybody gets pulled out in the spring when the blueberries are starting to break bud. The sheep are then rotated around all of our headland spaces so that we don't have to mow. They've significantly reduced the amount of time we spend on the mower. Integrating the animals has been really fun and it works for me.

Seasonal Life Cycles

In addition to the biogeochemical cycles described in the previous section, all the organisms on the farm have their own natural life cycles: crops, livestock, weeds, insects, and microorganisms. Insects and microorganisms can be crop pests or beneficial organisms that support your farm ecosystem, and sometimes both. In the following sections, we look at some of these in more detail. Keep in mind that other life on the farm includes natural vegetation, wildlife, arthropods, mollusks, and huge numbers of beneficial soil-dwelling organisms such as bacteria, fungi, nematodes, and other soil fauna. We do not address these in this chapter, but these organisms have their own natural life cycles that play out in the farm ecosystem.

CROPS

The basic life cycle of crop plants includes the following phases:

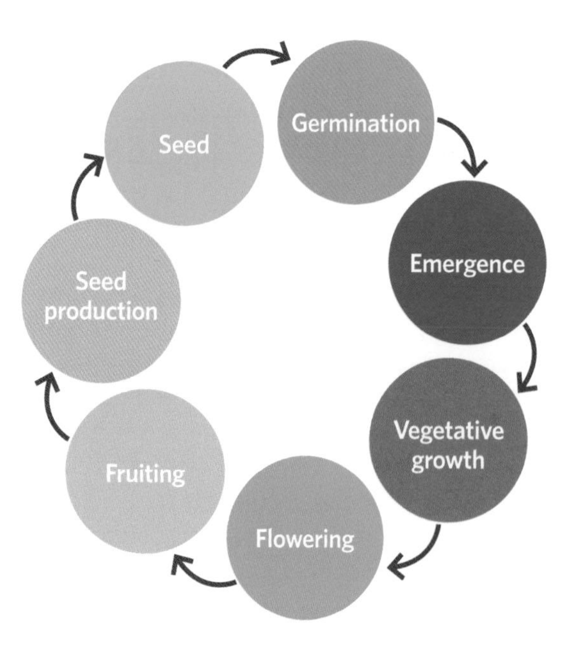

There are variations on this sequence (for example, some plants do not produce flowers or seeds), but this is the basic framework. Every crop has its own unique life cycle that is governed by genetics and environmental conditions (primarily temperature, moisture, and day length). Different crops are harvested at different developmental stages depending upon what part of the plant is typically eaten or utilized (for example, compare asparagus, spinach, broccoli, tomato, and winter squash).

The growth rate of a crop plant — the time in which it passes through its life cycle — may be managed by cultivar selection and the planting date. Crops are subdivided into three groups based on their lifespan and the timing of their reproductive phase:

Annual crops complete their life cycle once in a growing season, during which the vegetative and reproductive stages are completed (seed to flower to seed). Annual crops may be grown for their leaves (spinach, lettuce), fruit (peppers, tomatoes), or seeds (grains, legumes).

Biennial crops complete their life cycle in two growing seasons. During the first season, they produce roots and leaves. During the second season, they transition to a reproductive phase where they produce flowers, fruits, and seeds. Some biennial crops, such as carrots and onions, are often treated as annuals and grown for their vegetative stage only.

Perennial crops have long lives (technically more than two years). When mature, they produce a crop during each growing season. A number of perennial grasses and legumes are used for grazing and hay production. For woody perennials — such as vines, tree fruits, and nuts — there is generally a period of development, as the young plant grows to a size capable of producing a commercial-scale crop.

LIVESTOCK

Animal life cycles also follow a common sequence:

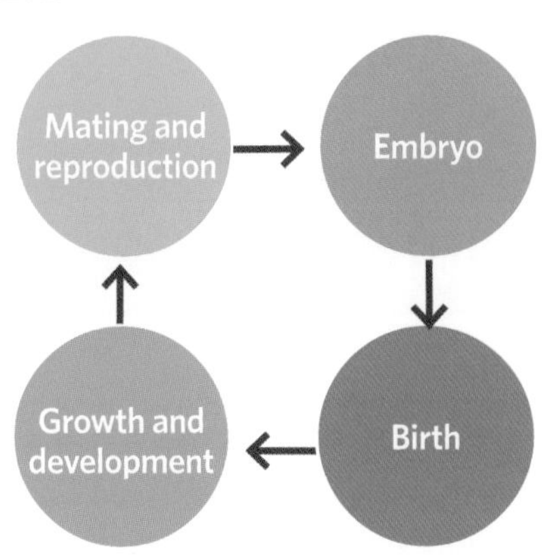

The life cycle for livestock grown for food begins at birth and ends at an intended point of harvest or when the animal is no longer productive for its intended use. Each animal species has its own unique life cycle, which farmers manage in order to produce a marketable product such as milk, meat, eggs, or wool.

PESTS

Pest organisms — weeds, insects, and diseases — add to the complexity of the farm ecosystem. Having a basic understanding of the biology and life cycles of the pest organisms within your farm ecosystem will help you manage them more effectively. In this section, the goal is to provide an overview of basic biology and how that plays out in the context of the farm ecosystem. We cover some specific pest management strategies in the Key Practices in Sustainable Agriculture section later in this chapter.

WEEDS

To begin with, what is a weed? One common definition is that weeds are plants that are growing where they are not wanted. Plants may be considered weeds in one context but not in another. For example, some gardeners and farmers consider medicinal herbs (for example, chicory, plantain, and mullein) to be weeds, while others are growing these same species as crops for the herbal market. Another example is yellow star thistle: while most farmers think of this plant as a scourge, beekeepers view the plant as good forage for producing honey.

Weeds pose many challenges. They compete with pasture and crop plants for soil nutrients, water, and sunlight, particularly during the crop's early growth stages. Even later in the season, weeds may compromise the health and yields of crops. And weeds with burrs, thorns, and needles are a nuisance for livestock and humans.

The same basic sequence of development that we looked at for crop plants applies to weeds.

Seed
Germination
Emergence
Vegetative growth
Flowering
Fruiting
Seed production

Knowing the life cycles of the weeds on your farm can help you determine when and how to control them. Weeds can be classified under five general categories: summer annuals, winter annuals, biennials, perennials, and creeping perennials.

REDROOT PIGWEED

HIMALAYAN BLACKBERRY

YELLOW STAR THISTLE

Summer annuals. These weeds germinate as soil temperatures rise in spring. They develop seed heads in summer or fall and complete their life cycle within one year. Some examples of summer annuals are redroot pigweed, common lamb's quarters, green foxtail, and Russian thistle.

Winter annuals. These weeds germinate anytime between fall and early spring. They develop seed heads during spring and also complete their life cycle within one year. Some examples are yellow star thistle, chickweed, downy brome, and shepherd's purse.

Biennials. These weeds complete their life cycle in two years. During the first season, they often produce a rosette — a circular arrangement of leaves — that grows near the ground. During the late summer or fall of the second year, the plant sends up a seed head from the rosette. Some examples are wild carrot, common mullein, and musk thistle.

Perennials. These weeds live for multiple seasons. Some examples are johnsongrass, buckhorn plantain, Himalayan blackberry, and kudzu.

Creeping perennials. In addition to reproducing by seed, these perennial weeds reproduce asexually through rhizomes and stolons. Some examples are field bindweed, quack grass, and Canada thistle.

Like all plants, weeds reproduce either from seed or vegetatively from plant parts such as rhizomes or stolons. For weeds that reproduce from seed, keep in mind that there is a seed bank in your soil. The seed bank represents all the viable weed seeds left over from previous years in which weeds reproduced in that field. Given the reproductive potential of most weeds, even a few seasons without weed control can result in a very large seed bank. Seeds can remain viable for years, so successful weed management is a multiyear process. As the farm ecosystem manager, your goal is to keep annual weeds from setting seed in the first place (this is part of planning in the planning-intervention framework). Control can be challenging in certain difficult-to-weed crops such as winter squash. In those situations, consider rotating with other crops where weed control is easier, such as lettuce or broccoli, and avoid following a weedy crop with a weed-sensitive crop like carrots.

For perennial weeds that reproduce vegetatively, intermittent cultivation often results in an increase in your weed problem, so be careful. Integrate complementary weed management strategies such as diligent cultivation, crop rotation, and cover crops rather than relying on one technique. Become acquainted with the life cycle of weeds that affect your farm so you can decide how — and at what life cycle stage — to control them. One place to start is the USDA Natural Resource Conservation Service's Introduced, Invasive, and Noxious Plants, which includes a list of noxious plants for each state in the United States (see Resources).

INSECTS

In scientific terminology, insects (Insecta), spiders and mites (Arachnida), and garden centipedes (Symphyla) are three separate classes of organisms within the larger grouping (phylum) known as Arthropoda. In this chapter, we use the term "insects" to refer to a number of different arthropod pests. Insects are an important part of farm ecosystems. Beneficial insects include pollinators, scavengers, and natural enemies of pests. Pest insects cause damage to crops and transmit diseases. As you learn how to identify and manage beneficial and pest insects on your farm, your farm ecosystem will become more efficient and robust.

Insect development. Metamorphosis is the change in form that insects undergo as they grow from eggs to adults. The two main types of metamorphosis are simple (or gradual) and complete.

Simple metamorphosis is the most basic form (see the illustration below). The egg hatches and the nymph emerges. The nymph feeds, grows, and molts, shedding its outer skin several times before reaching the adult stage. The immature stages of these insects look very similar to the adult stage, but the immature insects are smaller, may have fewer legs, and may be a different color. Adults of some species develop wings while their nymphs do not. Grasshoppers, true bugs, leafhoppers, and aphids are examples of insects that undergo simple metamorphosis. Aphids are a special case: they often do not lay eggs, but rather the embryos develop inside the female adult, and, when ready, she gives birth to nymphs.

SIMPLE (GRADUAL) METAMORPHOSIS

Generalized life cycle for insect undergoing simple metamorphosis

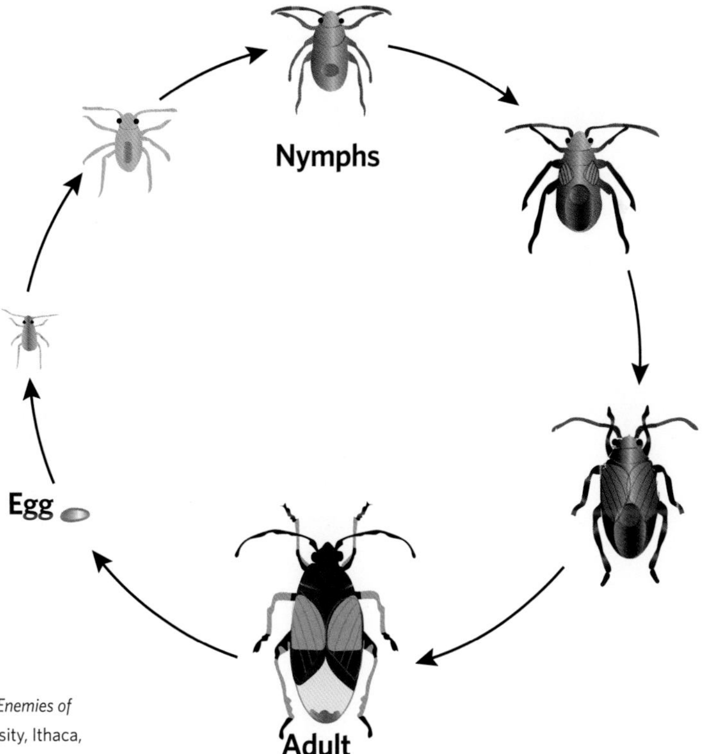

Source: Hoffmann, M.P. And Frodsham, A.C. (1993) *Natural Enemies of Vegetable Insect Pests*. Cooperative Extension, Cornell University, Ithaca, NY. 63 pp. / https://biocontrol.entomology.cornell.edu/bio.php

Complete metamorphosis is more complex, and the insects undergo drastic physical changes (see the illustration below). Larvae hatch from eggs, feed, then pupate. During pupation, they develop into fully formed adults (like butterflies). Adults emerge from the pupal case a little pale and soft, but within a few hours they are ready to take flight, look for adult food, meet their mate, and find new egg-laying sites. The larval stage of pest insects is often what causes the most crop damage.

Insects can have one or multiple generations per year, depending on climate and weather. As long as conditions are favorable, many species will continue reproducing. Other insects, like wireworms, can take more than one year to complete one generation. Insects can overwinter as eggs, nymphs, larvae, pupae, or adults. Many insects will spend the winter in a state of dormancy, called diapause, waiting for an environmental or genetic trigger to become active. During diapause, growth, differentiation, and metamorphosis stop. This resting state allows the insect to survive unfavorable conditions such as extreme low or high temperatures. Diapause is often broken by light, temperature, humidity, or physical disturbance.

Eggs vary in size, shape, and color depending on the species. Most eggs are spherical, oval, or elongate. It can take a few days to a few months for an egg to fully develop and hatch, depending on species and time of year.

Nymphs are the immature stages of insects with simple metamorphosis. Nymphs look very similar to the adult form and feed and live in similar habitats. They often coexist with adults in colonies on plants. Each stage of the nymph is called an instar, and the number of instars required before reaching sexual maturity varies depending on insect species.

COMPLETE METAMORPHOSIS

Generalized life cycle for insect undergoing complete metamorphosis

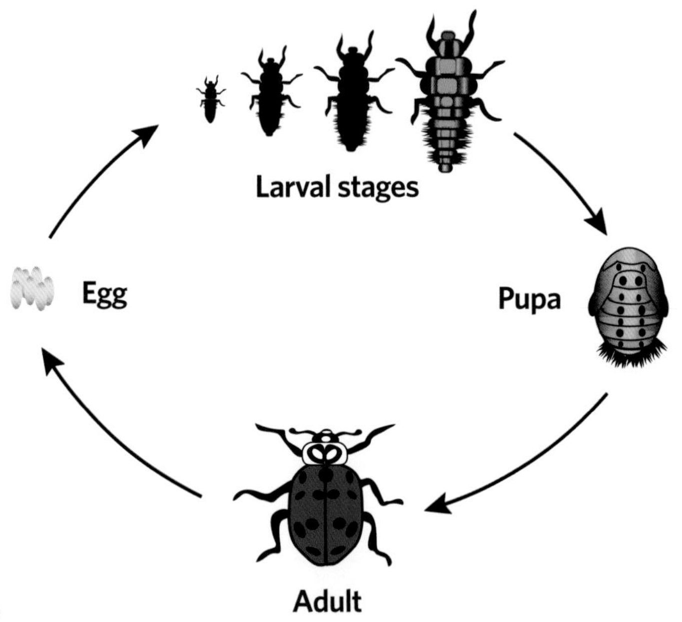

Source: Hoffmann, M.P. And Frodsham, A.C. (1993) *Natural Enemies of Vegetable Insect Pests.* Cooperative Extension, Cornell University, Ithaca, NY. 63 pp. / https://biocontrol.entomology.cornell.edu/bio.php

Larvae are the immature stage of insects with complete metamorphosis. Larvae come in several forms that impact where they feed and how mobile they are. The larvae of ladybugs and other beetles are predators of other insects, and they can move relatively quickly. Caterpillars move more slowly and feed on vegetation and fruit. You can often distinguish predator from pest larvae by their mobility and aggressive mouthparts. Some larvae do not have legs and are blind and immobile, completing their development inside plant parts, host insects, or colonies of prey; these include Diptera (fly) larvae and many of the Hymenoptera (bees, wasps, and ants).

Pupae occur when insects are undergoing complete metamorphosis. The pupal stage is when larvae are transformed into the adult form. It may appear that not much is happening in the pupal case, but complex biochemical changes are taking place. The insect is enclosed in a pupal case and many have an additional outer case or cocoon protecting it.

Adults have a primary purpose of reproducing. The adult stage is sexually mature, and many are able to fly, allowing them to travel to find a mate. Adults frequently have different food sources than the larvae, such as pollen and nectar. Once successful mating occurs, the female deposits eggs either singly or in clusters on or in plant tissue, soil, or other insects. Thus the cycle begins again.

Crop damage. The type of damage caused by an insect pest is usually related to the type of mouthparts it has. Even if an insect pest is not visible, you can sometimes identify it (and what stage of its life cycle it is in) by the kind of damage you are seeing on your crop.

Chewing mouthparts are possessed by caterpillars and many other larvae. At this stage in their life cycle, they do the most feeding and damage. Other insects with chewing mouthparts include grasshoppers, earwigs, and many beetles.

Sucking mouthparts are able to pierce through the cell wall and suck sap through the plant tissue. Aphids, scales, true bugs, and leafhoppers are in this group.

Rasping mouthparts slash the plant tissue to make a wound and then suck up sap oozing from the wound. Pests in this group include thrips, mites, and slugs.

Nematodes. Plant parasitic nematodes (also known as roundworms) range in length from 250 microns to 12 millimeters. They do not have a skeleton or a defined respiratory or circulatory system. They feed using a hollow mouth spear or stylet to pierce plant tissues, then they consume the contents of individual cells.

Nematodes begin their life cycle as an egg and progress through four juvenile stages to adult, shedding their cuticle between each juvenile stage. The first molt often occurs in the egg, and they hatch as second-stage juveniles. Although they are mobile, most nematodes move less than a yard through the soil in their lifetime. Many nematodes have environmentally resistant resting phases that allow them to survive dry or cold periods. Active or dormant nematodes can be transported with dust, soil, and plant material by wind, wild animals, and human activity.

Cyst- and root knot–type nematodes cause the most crop damage. They are completely embedded in the root for most of their life cycle and cause the plant to form permanent feeding cells. When female cyst nematodes die, their bodies create an environmentally resistant cyst around the eggs, allowing them to survive in soil for several years until cues from new host plants stimulate the eggs to hatch. Root knot nematodes are able to feed on a wide range of host plants to ensure their survival.

Roundworms are an example of a nematode that affects livestock. Adult nematodes produce eggs in the host animal, and the eggs are expelled from the host with the feces, contaminating the pasture. Larva hatch from the eggs and molt two times, becoming a third-stage larva that is capable of migrating from dung pats and soil onto moist grass. Infection occurs when the larva is consumed with the grass. Larvae can survive up to a year on pasture. Good pasture-management methods can disrupt this cycle.

As you develop an insect pest management strategy, keep in mind that your first goal is prevention: plan and design your system to avoid or minimize pest problems. Also monitor your crops. An effective pest management plan is based on knowing which pests you are dealing with, and at what point in their life cycle they are most susceptible to control. It is also important to recognize (and encourage) the beneficial insects on your farm.

DISEASES

Three conditions are needed for plant diseases to develop: a susceptible host, presence of the pathogen, and environmental conditions favorable to pathogen development. If any one of these conditions is missing, disease will not develop. The plant disease triangle gives us a way of looking at how these three factors interact (see the illustration below).

PLANT DISEASE TRIANGLE

The Plant Disease Triangle shows the conditions needed to make a plant susceptible to disease. If your management can interrupt one or more of these conditions, you can reduce disease pressure.

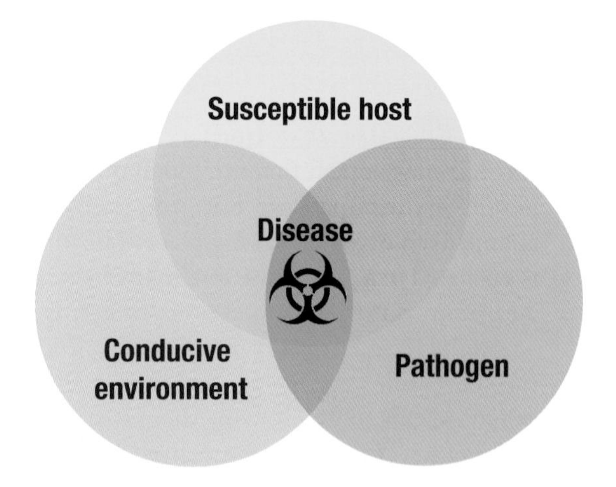

RELATIONSHIP BETWEEN MANAGEMENT PRACTICES AND PLANT DISEASE TRIANGLE FACTORS

MANAGEMENT PRACTICE	DISEASE TRIANGLE FACTOR AFFECTED
Grow downy mildew-resistant lettuce varieties	No susceptible host
Destroy potato cull piles to reduce late blight inoculum	Removes or reduces the amount of pathogen
Use drip irrigation to maintain drier leaf canopy and reduce the risk of foliar disease	Changes the environment to discourage disease development

The plant disease triangle also gives a clue as to how we might manage plant diseases. We can often modify one or more of the three factors — host, pathogen, or environment — to prevent the disease, or at least limit its development so it does not harm the crop. The table above shows three disease management practices, along with the part of the plant disease triangle that is affected.

Plant diseases originate from one of five different types of organisms: fungi, oomycetes, bacteria, viruses, and nematodes (discussed previously). These organisms are described in general terms. Note that each of the five types of organisms includes thousands of different species, each with their own particular characteristics.

Fungi. Most plant diseases are caused by fungi. Fungi obtain their food from other living organisms as decomposers, parasites, or pathogens. Scientists estimate there are at least 1.5 million fungal species on earth. Fewer than 10 percent of known fungi can colonize living plants; most of them are symbiotic (mycorrhiza and chanterelles) or decomposers (brown and white wood-rotting fungi). Many soil-living species are important for developing soil structure; other soil fungi are capable of strangling plant-parasitic nematodes. Plant-parasitic fungi obtain food from their host, but the host plant doesn't necessarily exhibit symptoms; plant-pathogenic fungi cause disease usually by interfering with metabolic functions of the crop.

Oomycetes. Oomycetes, once thought to be a type of fungi, have their own distinct characteristics and are now classified separately. Common pathogens in this category include "water molds" such as *Phytophthora infestans* (late blight of potatoes and tomatoes), *Plasmopara viticola* (grapevine downy mildew), and *Pythium ultimum* (seed rot and damping-off).

Bacteria. Bacteria are single-celled microorganisms. All plant surfaces have bacteria on them, and various other bacteria live inside plants. They can be beneficial or detrimental to the plant. Some disease-causing bacteria organize in dense colonies to form biofilms that protect them from harsh environmental conditions and help them attach to plant surfaces. Environmental conditions have to be just right for this kind of proliferation to occur.

Most bacteria are aerobic, surviving in plant debris left on the soil surface, in soil, in seeds, and in perennial hosts. Bacteria move passively in the environment and are disseminated with windblown soil and in water. Plant infection is also generally passive: bacteria can be sucked into plants via leaf openings

DISEASE PEST IDENTIFICATION AND MANAGEMENT RESOURCES

When managing specific diseases, refer to reference materials or ask the advice of a specialist to correctly identify the disease and the stage of its life cycle during which controls are most effective. Prioritize your learning according to the diseases you are most likely to face on your crops and in your farm environment. Cooperative Extension Service resources can help you identify where to place your focus.

(e.g., stomata), through wounds on the plant surface, or by insects feeding on the host plant.

Viruses. Viruses need a host plant to complete their life cycle. Viruses consist of a nucleic acid with a protective protein shell, and they rely on their host's RNA for energy and to reproduce. Outside their host, viruses are immobile and rely on other organisms (vectors) or the environment to disseminate and infect host plants.

Most viruses are transmitted from infected plants to healthy plants by insects, nematodes, or fungi. Aphids and whiteflies transmit the most plant viruses. Viruses are also transmitted through vegetative propagation (as with potatoes) or seed. Some viruses are carried by insect vectors on their mouthparts and transmit within seconds to minutes, but they do not survive long on the vector. Other viruses move through the insect vector's gut wall and multiply and circulate throughout the insect and are then transmitted through their saliva. These viruses are spread by that insect for the rest of its life.

Plant and animal disease diagnosis is challenging, so you will need the help of a diagnostician. You can check with your local Cooperative Extension Service. In addition, many land-grant universities have plant health or veterinary clinics you can take or send samples to. Correct identification is the critical first step in managing diseases.

MANAGING THE FARM ECOSYSTEM

At the beginning of this chapter, we used Eliot Coleman's analogy of the farmer or rancher as orchestra conductor to illustrate the farmer's or rancher's role as an ecosystem manager. Using that same analogy, the ecosystem cycles described above can be thought of as different sections of the orchestra.

Now that you have been introduced to the biogeochemical and seasonal life cycles on a farm, we are going to bring those sections of the orchestra back together to perform. In this section, we address how to account for and integrate these different cycles in a way that helps you reach your production and economic goals.

Farmers Share Their Experience

ANNUAL VEGETABLE PRODUCTION IN SOUTHERN OREGON

MELANIE KUEGLER AND CHRIS JAGGER share how their awareness of the different cycles at work on their farm have helped them become better farm managers. Note how they continually assess the farm ecosystem, identify where there are problems and challenges, and implement practices that keep the farm system productive and in balance. What are some of the key questions that will help you assess the overall health of your own farm ecosystem?

CHRIS JAGGER AND MELANIE KUEGLER, BLUE FOX FARM (profile on page 168)

Chris: What we are looking for in a property has definitely changed. When we bought our home property in 2003, we had one mind-set: we wanted a property that had some elevation change so we could have a gravity-fed irrigation system. We were also looking for something that could handle a lot of animals as well as vegetables because we were looking for a more polycultural farm.

When we bought our other piece of land in 2007, the one that we're now actively farming on, we had a completely different perspective. We were looking for flat land with good soil and good irrigation, period.

It gets really hot down here in southern Oregon. We easily hit temperatures above 100°F (38°C), and it's very dry. We thought the climate would restrict what we could grow, ruling out greens and lettuces and things like that, but we found systems that could work around the heat restrictions. We probably don't have the most ideal climate to grow spring mix in the middle of summer, but we figured out ways to pull it off.

Melanie: It's funny because things that can be a limitation can also be a benefit. We don't have the same downy mildew and disease problems that they have where the climate is milder but wetter. So we have enough water and good enough soil that we're able to overcome a lot of the limitations that you see with hot and dry climates. It's cold in the winter, too. We're not in California. You can't grow through the winter, but we've overcome that with storage.

Row covers and hoop houses can extend the growing season by weeks.

Season Extension

Melanie: We learned a lot about season extension when we grew in Colorado, where we were growing at a mile high. We've applied that knowledge here.

We have seven hoop houses, put row cover out in the field, and store stuff. These probably extend the season by a few weeks in spring, and they help us in fall. In any given season, those few weeks are pretty critical in terms of making income off of a crop. We're able to do farmers' markets from the first week of March through Thanksgiving.

Chris: We do a lot less with drip tape than we thought we would. We discovered how much more efficient overhead water is at cooling down the fields in the heat of the summer. People are always amazed that we'll have lettuce at market when it's 103°F (39°C). With drip tape, you just can't get the soil temp down.

Melanie: We have a barn with a lot of storage capability. We've learned a lot about pulling stuff out of the ground early, storing it through winter, and having it ready in March, when we start selling at markets again. We also wholesale that stuff through winter.

Maintaining Soil Health

Melanie: We cover crop all of our property at some point during the season. We usually give a piece of land a two-year rest with some sort of perennial cover. Then we rotate our crops in spring, summer, and fall. The land is planted in spring, then rested in summer, and put into an early fall cover that goes through the season. Only one piece of land is in crop production in winter.

We also add a pelleted fertilizer for nutrients at the time of planting, and we add trace minerals as needed, which is determined by soil testing.

Chris: We do use a rototiller a lot. We looked into using a spader instead, but it's so much slower and it seems like everybody I know who has a spader is always fixing it. We also use a Yeomans plow for deep subsoilage, to break up any plow pan. The rotation that we have — resting the ground and subsoiling before we put that area

into a fallow, which allows any of that compaction to loosen up over the time — has been our primary way of dealing with tilth.

We have moved toward not overworking pieces of land and really paying attention to what the soil strata looks like. You can learn a lot by observing what weeds are growing. If you see a lot of taproot start showing up you, the soil's asking to be opened up again.

Pest and Disease Management

Melanie: I think our cover crop rotation is the biggest thing we've done to limit pests and diseases. We used to grow on a small amount of property, filling it out in spring, tilling beds as they're done, throwing something else in there, tilling that as it's done, then throwing something else in there. Every fall we would be covered in pests, and we'd start seeing diseases pop up. We don't have those problems now.

We have a spring area that is planted, then put into cover crop after the crops are harvested. It's not used again until the next year. As a result, we have gorgeous fall crops right now, free of aphids, which is something that we have not had before. We've battled aphids every single year prior, but now we're seeing very few of them. Row cover has also been a godsend. We use it not only for season extension but also to keep pests, particularly flea beetles, off of our crops.

We've used *Bacillus thuringiensis* (Bt) or organic pyrethrum in extreme cases, but we try to keep that at a minimum, using it only in cases where we can see that something has become an epidemic. But we also investigate why the epidemic popped up and try to make changes to our system so that it doesn't happen again.

Chris: We've just stopped growing certain crops because every year they were enough of a problem that it wasn't worth it for us to have to continually invest in inputs to deal with it. That's just not for us.

Guiding Principles

The health of a farm ecosystem is part of whole farm planning. We introduced whole farm planning in chapter 1, "Dream It," and you have had the opportunity to work on different components of your plan in the farm plan worksheets presented in this book. From this holistic perspective, you can use three overarching principles to guide your crop and livestock management decisions.

1. Choose crops and livestock that are well adapted to your area and create an optimal growing environment for those crops and livestock.
2. Promote soil quality and health.
3. Enhance biodiversity on your farm.

As we address each of these in turn, note how these principles relate to the various ecosystem cycles described previously. Also note how these principles fit within the planning and intervention framework presented earlier in this chapter and translate into specific production practices you can apply on your farm or ranch. We explore some of those practices in the Key Practices in Sustainable Agriculture section later in this chapter.

1. CHOOSE CROPS AND LIVESTOCK THAT ARE ADAPTED TO YOUR AREA

This first management principle states simply that you need to know the specific requirements for the crops and/or livestock you are interested in producing, evaluate their adaptability to your farm environment, then design and manage a farming system that provides essential resources efficiently. That means providing the quantities needed, at the right time, and with least disruption to the environment (minimizing losses).

The aim is to grow crops — or raise livestock — that are healthy and strong and that can withstand environmental extremes

and pest pressures. Part of that process involves understanding what the land is able to support, what crops or livestock species are compatible with your location and resources, and the timing of the seasonal cycles on your farm. Let's look at each of these factors more closely.

Land resources. Important considerations include sun exposure, soil type (texture, pH, and depth), water availability, aspect, slope, air drainage, wind, climate, and access. Additional considerations for livestock include proximity to adequate water supply, access to land for manure application, adequate space for social and reproductive needs, distance from neighbors, and shelter. These are all important topics addressed in chapter 1, "Dream It."

Compatibility. The compatibility of crop and livestock species with your particular farm is primarily a matter of genetics. Plant and livestock species have evolved and adapted to certain climates, so selecting those that are compatible with your conditions will make your job a bit easier and less stressful. For example, Kiko goats tend to have fewer foot and parasite problems in wet climates compared to other goat breeds. In cool, wet climates, tomato and potato varieties with resistance to early blight and late blight usually perform better than susceptible varieties. Also, be aware of the expectations of the market you are producing for. Some markets are more accepting of unique varieties or breeds, whereas others prefer standard products.

Timing. The yield, quality, and marketability of your product are all affected by its development through the growing season, so timing is key. Soil must be cultivated and amended at the right time to minimize compaction and nutrient leaching. Crops must be planted at the right time to ensure the best growing conditions. Pest management activities also have to be well timed. Additionally, timing is important when you want to sell your product during a window of time that allows you to get a better price. Season-extension techniques are often used to support this strategy. If you are a livestock producer, timing is a critical factor in making decisions about feed (for example, pasture or hay), water, and shelter. Timing of breeding, calving, and harvesting are important, too, as these activities affect when you have product available to sell.

2. PROMOTE SOIL QUALITY AND HEALTH

Soil is the foundation of every farm, and good soil management practices maintain or improve soil health and quality. We use the terms *soil health* and *soil quality* interchangeably, as both refer to the soil's role of supporting the growth and yield of crops. Characteristics of a healthy soil include:

- Sufficient nutrients
- Good structure with strong aggregates that resist degradation
- Good water infiltration
- Good water drainage
- Good water-holding capacity
- Low populations of pest organisms and weed seeds
- Healthy populations of beneficial soil organisms (such as ground beetles, earthworms, springtails, beneficial nematodes, bacteria, and fungi)
- No or minimal harmful chemicals
- Rich, earthy smell
- Plant cover throughout the year
- Surface that does not easily crust
- Robust soil food web
- High organic matter content

In the field, a variety of indicators will give you a sense of the quality of a soil, such as texture, porosity, aggregation, compaction, color, and smell. You can also use visual clues from the morphology and condition of plants growing in the soil, such as leaf color, size, and vigor of the plant. If you have completed the farm plan worksheets for chapter 1, "Dream It," then you have already done some work to assess the soils on your farm: you determined its texture and identified its capability class and major limitations that might affect your farming practices.

Healthy and productive soils contain organic matter, although it is generally present in relatively small amounts. Typical agricultural soils may have anywhere from 1 to 6 percent organic matter depending on location, soil type, and management. Soil organic matter consists of three general categories of materials: living organisms, fresh and partially decomposed residues, and well-decomposed residues (humus).

How can organic matter, which makes up only a small percentage of most soils, be so important? The reason is that organic matter positively influences essentially all soil properties. Soil organic matter:

- Improves soil fertility
- Increases nutrient retention (cation-exchange capacity)
- Improves soil tilth and aggregate stability
- Improves water infiltration and water-holding capacity
- Improves biological activity
- Promotes healthy plants
- Reduces soil erosion
- Reduces insect and disease problems
- Reduces allelopathic and phytotoxic effects

Many of the practices discussed in the final section of this chapter focus on promoting soil health and quality.

3. ENHANCE BIODIVERSITY ON THE FARM

The third principle for managing the farm ecosystem is based on the concept that an optimally diverse ecosystem needs fewer external inputs, capitalizes on the ecological processes that support the functioning of the farm ecosystem, and moderates any negative environmental impacts from farming. Diverse agricultural systems can:

- Encourage positive biological interactions (symbiosis)
- Enhance the opportunities for beneficial organisms
- Lead to more efficient use of resources such as nutrients and water
- Reduce environmental and economic risks for the farmer
- Improve nutrient cycling on the farm/ranch, especially as livestock are integrated into the system

There are a number of ways to characterize the diversity of an ecosystem. Some aspects of ecosystem diversity include:

- Number of different species in the system
- Degree of genetic variability within and among species
- Number of distinct physical layers or levels
- Number of habitats and niches
- Diversity of the soil food web
- Complexity of interaction, energy flow, and nutrient cycling among the components of the ecosystem

Farm ecosystems are generally much less diverse overall than natural systems. Note that many of the practices discussed next in the final section of this chapter (such as cover crops, crop rotation, and farmscaping) contribute to farm ecosystem diversity.

KEY PRACTICES IN SUSTAINABLE AGRICULTURE

Given the diversity of farming and ranching systems across the country, we cannot cover every type of crop or livestock system, nor can we address all the specific production strategies that farmers use to manage their farm ecosystem. Therefore, our aim in this final section is to provide an overview of some of the key practices for managing sustainable farming and ranching systems. Note that these key practices are all, in some manner, tied to the natural cycles discussed previously (most exhibiting aspects of energy flow and the carbon, nutrient, and water cycles). They work for the long-term health and resiliency of the farm or ranch and are utilized to disrupt the life cycles of seasonal pests. Also, keep in mind that the practices highlighted here have been modified and adapted in thousands of ways to fit the context and needs of different farm systems across the country. Key practices in this section are adapted by permission from Agricultural Ecosystem Management, Course 3 in the National Continuing Education Program, Sustainable Agriculture Research and Education (SARE), USDA National Institute of Food and Agriculture (NIFA).

Cover Crops

Cover crops provide numerous benefits for the farm ecosystem. They conserve soil — the roots hold soil in place and foliage protects it from the direct impact of raindrops, thereby reducing compaction and erosion. In addition, cover crops:

- **Reduce fertilizer costs.** Legume cover crops can contribute a significant amount of nitrogen and other nutrients to the soil. Cereals can scavenge residual soil nitrates left over at the end of the season.
- **Reduce weeds and other pests.** Cover crops can outcompete weeds for water, nutrients, and space; reduce soilborne diseases; provide habitat, pollen, and nectar for pollinators and other beneficial insects; and produce compounds that reduce some nematode pest populations.

- **Increase soil biodiversity.** Cover crop roots provide food and habitat for beneficial soil organisms such as mycorrhizal fungi and other beneficial microorganisms.
- **Enhance soil organic matter.** Cover crops can help maintain or improve soil organic matter levels in the soil.
- **Conserve soil moisture.** The added organic matter from cover crops can help increase water infiltration and reduce evaporation.
- **Protect water quality.** By slowing runoff and reducing erosion, cover crops reduce nonpoint source water pollution caused by sediments and nutrients. By storing excess soil nitrogen, cover crops prevent nitrogen from leaching into groundwater.

Cover crops can be annual, biennial, or perennial plants and can be grown as single species or diverse mixes. Farmers often select particular cover crop species to solve specific problems such as weed control, increasing soil organic matter, or adding fertility. Some cover crops also have unexpected farm cash flow benefits. For instance, in cover crop mixes that include field peas, the pea tendrils may be harvested as an early-season marketable crop prior to the cover crop being incorporated.

With intensive vegetable rotations, you often have a limited window of time for planting cover crops. For this reason, it is important to plan ahead and have the seed and inoculum (for legume cover crops) on hand, so you can seed cover crops when the opportunity arises. If broadcasting seed, use higher seeding rates than you would use with a drill, and be sure to bury large seeded cover crops (for example, using a ring roller or seed drill). Plan ahead so you can incorporate the cover crop about one month before planting your cash crop.

There are several online decision tools that can help you select the right cover crops for your farm (see Resources):

- Organic Fertilizer and Cover Crop Calculators from the Oregon State University Extension Service
- Cover Crop Guide for New York Vegetable Growers from Cornell University
- Cover Crop Decision Tools from the Midwest Cover Crops Council

Crop Rotation

Crop rotation is the primary tool farmers use to diversify their farming enterprises. Crop rotations on diversified farms are complex. The type and sequence of crops will vary based on the evolving business and management goals of the farm and the unique soils, water availability, and microclimate of individual fields. Rotations can include annual, perennial, and feed crops, plus cover crops and fallow periods. In planning your crop rotation, consider what goals you are working toward: solving pest or soil problems, controlling weeds, building soil organic matter, enhancing populations of beneficial organisms, reducing economic risks, increasing farm biodiversity and number of marketable crops, or other goals.

Good record keeping is an important part of managing crop rotations. Creating a farm map specifically for rotations is also helpful. In addition to showing the location of crops, the map should include notes for pest and soil problems. The maps from previous seasons provide guidance for where to locate future plantings. For annual crops, a sound rotation plan should aim to have the greatest number of years possible between similar crops in the same field. You may be able to keep track of cover crops and applications of organic amendments on these maps, too. Well-designed crop rotations can:

- Maintain soil health and quality due to variations in rooting depths, tillage practices, and nutrient needs
- Reduce weed, insect, nematode, and disease problems (study pest biology to learn whether rotation can help reduce pest populations or crop damage)
- Contribute to nutrient management plans
- Increase profitability over the long term through market diversification
- Enhance ecological diversity
- Provide year-round employment for farm labor
- Help minimize off-farm inputs

To learn more about crop rotation, check out *Crop Rotation on Organic Farms: A Planning Manual* (see Resources).

ORGANIC MIXED VEGETABLE FARM ROTATION (NORTHEAST UNITED STATES)

Potatoes, summer squash, root crops (carrots, turnips, et cetera), beans, tomatoes, peas, brassicas (cabbage, broccoli, and so on), and sweet corn are grown on 1½ acres, divided into eight subplots planted in these sequences:

Potatoes » squash: Both summer squash and potatoes can be kept weed-free relatively easily. Thus, there will be fewer weeds to contend with in the next rotation, the root crops, which are among the most difficult to keep cleanly cultivated.

Squash » root crops: Having two "cleaning" crops back-to-back prior to root crops can reduce weed problems in the root crops.

Root crops » beans: Beans seem to be least affected by detrimental soil effects that certain root crops such as carrots and beets may cause the following year.

Beans » tomatoes: This sequence separates tomatoes and potatoes (related species) by four years; this is important for pest control.

Tomatoes » peas: Peas need an early seed bed; tomatoes are undersown with a non-winter-hardy green manure crop that provides soil protection but doesn't cause decomposition or regrowth problems the following spring, when peas are planted.

Peas » brassicas: This sequence works out well in terms of timing for harvest and field preparation. The pea crop is finished and the ground is cleared early, allowing a vigorous green manure crop to be established by early fall.

Brassicas » sweet corn: In contrast to many other crops, corn shows no yield decline when following a crop of brassicas. Also, many brassicas can be undersown to a leguminous green manure which, when incorporated the following spring, provides ideal growing conditions for sweet corn.

Corn » potatoes: Research has shown corn to be one of the preceding crops that most benefit the yield of potatoes.

Adapted from *The New Organic Grower* by Eliot Coleman (Chelsea Green, 1989).

Soil pH

The acidity or alkalinity (pH) in a field is determined largely by the soil's parent material, and it is influenced by climate (such as rainfall) and management practices (such as soil amendments). Some soils are naturally acidic with a pH below 6 (Western Oregon and Washington, for instance) and some soils are naturally alkaline with a pH above 7 (many soils in central and eastern Oregon and Washington, for example).

Soil pH can affect nutrient availability and disease susceptibility in cropping systems. Some crops, like blueberries and many nursery crops such as azaleas and rhododendron, prefer acidic soil, although they may tolerate neutral or slightly alkaline soils. Most crops prefer a relatively neutral pH (6.0–7.0). Legumes and some other crops like spinach are particularly sensitive to low soil pH.

You can monitor soil pH every few years during routine soil testing. In addition to soil pH analysis, be sure to request the "SMP buffer" or "lime requirement" test. Most crops benefit from occasional applications of agricultural limestone in naturally acidic soils. Acid-loving crops grown in neutral or alkaline soils may benefit from applications of elemental sulfur, which will reduce soil pH. University crop nutrient management guides can help you interpret soil test results and determine rates of application for agricultural limestone or sulfur.

Compost and Soil Amendments

Compost, organic fertilizers, cover crops, and other amendments (such as manure, crop residues, crop processing waste, and green waste) are commonly applied in sustainable farming systems. Unlike synthetic fertilizers that make nutrients immediately available to the plant — often in a short-lived burst — organic materials release nutrients that become available through the breakdown (cycling) of soil organic matter. They also influence plant growth indirectly, by improving the soil's physical condition. Be thoughtful when choosing organic amendments and managing and storing compost. Some organic amendments may be contaminated with and may introduce weed seeds, symphylans, or soilborne diseases, so be sure the materials you use are free of contaminants.

Composting is a controlled process. Materials to be composted (feedstock) are selected based on their carbon-to-nitrogen ratio and moisture content and are mixed to a specific ratio. Feedstock is mixed into piles, windrows, or aerated static piles. Well-built piles typically have a carbon-to-nitrogen (C:N) ratio from 25:1 to 40:1 and 40 to 60 percent moisture. Woody materials often must be chopped or ground to facilitate composting. The goal in composting is to achieve the optimum air, moisture, and C:N ratio to promote the growth of microorganisms that break down the raw organic materials. Heat is produced during this process, so you can monitor progress by measuring the temperature within the pile. Proper composting can eliminate pathogens and weed seeds that might have been present in the feedstock. Finished compost also has reduced volume and particle size, making the material easier to haul and apply.

The proper conditions for composting include the following:

- Oxygen for microbial respiration (Make sure the compost pile is properly aerated.)
- A moisture content between 40 and 65 percent (The initial moisture content can be adjusted by adding water or covering the pile to avoid drying out or leaching during the composting process.)
- Particle sizes of composting materials of approximately ⅛ inch to 2 inches in diameter
- A carbon-to-nitrogen (C:N) ratio between 25:1 and 40:1

- A temperature range of 130 to 150°F (54 to 66°C) during the active (thermophyllic) phase, if pathogen or weed seed reduction is the goal. Temperature is an important way to reduce human pathogens in feedstock such as manure.

Finished compost is spread and incorporated into fields before planting, or it is spread on the surface in perennial systems like orchards and pastures. Chemical analysis of the compost reveals its nutrient content, and application rates can then be matched to the needs of a specific crop or rotation.

Uncomposted organic materials are cheaper, but they are often bulkier, more difficult to apply, and less predictable in terms of their effect on soil. Also, weed seeds, pathogens, and other contaminants can be a problem, depending on the source. Pay attention to the C:N ratio of raw organic materials and how that may affect the availability of nitrogen for your growing crops. Very low nitrogen materials such as straw or wood products use nitrogen from the surrounding soil during the decomposition process. Uncomposted amendments with less than about 2 percent nitrogen dry weight basis (20:1 C:N ratio) immobilize soil nitrogen during the first few months of decomposition; higher-nitrogen materials (such as legume cover crops and some manures) release mineral nitrogen during decomposition. The table below shows the typical nitrogen content of various organic amendments.

Organic materials applied to the soil contain all of the macro- and micronutrients that plants need from soil, but not always in the right proportions. For example, the N-P-K (nitrogen-phosphorus-potassium) ratio required by vegetables is roughly 3-1-3, and the ratio of many commonly used manures and organic amendments is often about 1-1-1. Plant-available nitrogen (N) is only a fraction of total nitrogen. This makes it easy to overapply phosphorus (P) in order to meet nitrogen requirements. Lower-nitrogen materials tend to build soil organic matter more quickly, and higher-nitrogen materials supply rapidly available nitrogen for plant growth. Consider the types of organic materials that are readily available in your area and how you could use them to improve soil organic matter and crop production. If you have high soil phosphorus levels and are near surface water or have sloped fields, look for ways to supply enough nitrogen without overapplying phosphorus over the long term.

Reduced Tillage

Farmers use tillage to incorporate cover crops and other organic amendments, for weed control, and to create planting beds and seedbeds. These are important functions, but tillage can

TYPICAL NITROGEN CONTENT OF VARIOUS ORGANIC AMENDMENTS

LOW NITROGEN (>1%) IMMOBILIZE N	MEDIUM NITROGEN (1-2.5%)	HIGH NITROGEN (>2.5%) MINERALIZE N
Bark	Leaf mulch	Poultry manure
Sawdust	Nonlegume cover crops	Legume cover crops
Straw	Cow, horse, goat, sheep manure with bedding	Feather meal and blood meal
Wood chips	Compost	Fish meal

lead to soil problems, especially when done incorrectly or in excess. Tillage breaks down soil structure and can cause soil compaction; it also oxidizes carbon and can reduce soil organic matter levels. In sustainable farming systems, farmers aim to reduce tillage as much as possible and avoid tilling when the soil is too wet or too dry. Perennial cropping systems such as berries, orchards, pastures, and woodlots are at less risk, but compaction from management and harvest practices has been shown to reduce crop and soil health. The underlying concept of reduced-tillage practices is to conserve and manage soil organic matter and reduce the disruption of soil structure and ecology.

Reduced-tillage systems are based on the concept that tillage should be minimized and/or restricted to the planting zone and does not have to disturb the entire field. The terminology related to the various methods can be confusing.

- **Reduced tillage** generally refers to any tillage system that is less intensive and that employs fewer trips across the field than traditional tillage.
- **Conservation tillage** is an overarching term for any method of soil cultivation that leaves crop residue on the soil surface, rather than plowing it in.
- **Ridge-till systems** plant crops on permanent ridges about 4 to 6 inches high. The previous crop's residue is cleared off ridgetops into adjacent furrows at planting to create a good seedbed. It is essential that ridges are maintained, and this requires modified or specialized equipment.
- **No-till systems** leave the soil untilled throughout the growing season. Soil is disturbed only at planting by coulters or seed disk openers on seeders or drills, and weed control is often accomplished with herbicides.
- **Strip-tillage systems** are a variation of no-till in which just the seed row is tilled prior to planting to allow residue removal, soil drying, and warming in the planting zone.

Implementing these types of systems takes creativity and commitment. Whichever technique you choose, carefully weigh the benefits and problems of reduced tillage practices. Benefits include:

- Reduced erosion
- Improved soil quality
- Cost savings from reduced labor and energy inputs
- Increased availability of water for crop production
- Enhanced soil microorganism activity

Some of the key challenges of reduced-tillage practices include:

- Residue management, particularly related to planting of the subsequent cash crop (This is especially challenging if high-biomass cover crops are used in the rotation). Specialized equipment has been developed for the purpose of planting through the surface residue into untilled soil
- Cooler soil temperatures that can delay germination
- Other changes in the soil environment (for example, soil structure or soil moisture content) that may lead to pest problems
- Increased problems with slugs and rodents.
- Weed control challenges
- Potentially lower nitrogen and phosphorous levels in the soil may require revised nutrient management practices

Management-Intensive Grazing

Continuous grazing, when animals are allowed to graze wherever they want on a large pasture, is highly inefficient. Under these conditions, young tender plants (and forage near water, shade, or shelter) will be overgrazed, while more mature plants and distant portions of the pasture will be underutilized. This approach reduces animal production and changes the

MIG BENEFITS

- The intake of forage on a daily basis is controlled to meet livestock production needs.
- Pasture plants are allowed to recover after being grazed.
- Pasture yield is increased.
- The distribution of the forage is improved.
- The need for supplemental feed, equipment, and fuel is reduced.
- Animal waste is more uniformly distributed across the pasture, improving fertility and soil quality.

MIG CHALLENGES

- The system requires a lot of fencing.
- Water for livestock must be available at each paddock.
- Additional time is required for thoughtful management and moving animals.

plant community of the pasture, often shifting it toward less-productive species.

A management-intensive grazing (MIG) system closely controls forage production and harvest. A MIG system divides a pasture into a number of smaller pastures, called paddocks. Animals are allowed to graze the forage within a paddock and are moved to a new paddock when the forage is grazed to a specific height. Recently grazed paddocks are allowed to recover and regrow to an optimal height before being grazed again. The size and number of paddocks and the frequency of moves are linked to the physical resources of the farm, the goals of the manager, as well as the production needs of the livestock.

An important and often profitable approach to MIG livestock production is to graze multiple species over the same pasture. Different livestock species graze different portions of a pasture. What one type of livestock will not eat, another type will. Goats, sheep, cattle, plus broilers and laying hens can all be grazed using this approach. When well managed, this approach allows excellent pasture utilization and increased livestock production without overuse of the pasture. MIG systems have a number of important benefits, and a few trade-offs to consider (see table above).

FORAGE QUALITY

Knowing the nutrient composition of feeds and matching feeds to animal requirements at a given stage of production will help ensure that nutritional needs are met. Select pasture forage species that are adapted to your site conditions, livestock needs, and management style. Some forage is more palatable to certain species than to others. Since many factors affect forage quality, no single factor can be used to make this prediction. Maturity stage at harvest, forage species and variety, leafiness, and harvest and storage conditions are important factors that determine quality. From a systems perspective, it is also important to note that nutrients are coming out of the backside of livestock and, if managed properly, can be a great nutrient source for your crops and help you protect water quality.

Farmers Share Their Experience
LIVESTOCK PRODUCTION STRATEGIES

GOODFOOT FARM AND RAINSHADOW ORGANICS have integrated livestock and poultry into an annual production system, but at Sweet Home Farms, livestock and poultry were the main focus. As you read the comments from Carla Green and Mike Polen, note how the farm ecosystem was a constant reference point in management decisions for maintaining pasture condition, improving the productivity of the soil, protecting waterways on the farm, and dealing with predators. How did their system change over time, and what drove those changes?

CARLA GREEN AND MIKE POLEN, SWEET HOME FARMS (profile on page 216)

Mike: You can pretty much count on it being wet from November through June and then dry, very dry, in July, August, and September. By October, we usually want some rain. That affects the way we use pastures for different classes of animals.

The labor here varies tremendously by season. In late autumn, it's a transition time. We're not yet feeding grass hay to animals, so there's a relatively small amount of outdoor labor. We harvest a lot of beef and pigs and goats and sheep, so one of the main things we do is haul animals to the slaughterhouses, then bring meat back and figure out how to store it.

In spring, as we start grazing again, the work will change. We won't feed hay, but we will move cattle from paddock to paddock. So we'll be moving water troughs, moving electric polywire, and monitoring the grass to see that it's grazed as evenly as we can manage.

Summer work is pretty similar — we are moving animals between different properties and selling, of course.

We have two poultry enterprises. We have laying hens, which are year-round and raised on pasture. We have two houses yoked together that are pulled to a new section of the pasture every few days, depending on the weather. And we also have a seasonal broiler enterprise. We tend to lose too many broilers if we start in March, so we usually don't start raising them until May. The shelters are moved twice a day. It is pretty intense labor.

The one principle that seems to be consistent is that however much we plan, every day is different. There's almost always something unexpected that comes up that kind of slants your morning or your afternoon and takes several hours away from what you planned to do. This is why little things don't get done around the farm, and things pile up.

Carla: When we started doing the economic analyses on our different enterprises, we realized that we couldn't be profitable feeding hay to cattle through the winter. So for this climate, where you can't stockpile forage for the winter and where hay costs are high, we just couldn't figure out how to make it work. So we moved to a seasonal operation where we could finish animals on high-quality grass and butcher them before winter. That was an important evolution in our understanding of how to work with the local climate.

We have a grazing plan that shows what animals are supposed to go where at any particular time. That's the general plan, but it is altered depending on what happens with the weather.

We try to think things through, asking questions like "What's the carrying capacity of each field?" and "How many days are we going to have for each animal type in each field?"

We have both livestock-guarding dogs and herding dogs, which are essential to our farm. On our property, we feel comfortable having single dogs with the herd, but our leased properties are in areas of greater predator pressure, so we like to have pairs because there are cougars. Our herding dog — we have an English shepherd — is worth at least three people.

We have a system where we can feed animals in the barn during the wintertime and they all have access to pasture. We have feeders that are on pulleys that can be raised up as the bedding builds up. We continue to add straw throughout the winter to keep them warm and dry, then once they are out of the barn, we put the pigs in. The pigs root that up, helping to make compost. We also turn it a little bit with the tractor to facilitate the process.

We also have a compost barn. In fall, all that compost goes out onto the field as fertilizer and to build the soil. It is very, very effective. When we first came here, the farm had been continuously grazed. It had huge patches of bare and mossy ground, but that was all gone after we moved in. Between the management-intensive grazing that allows a rest and recovery period and the compost, the soils and forage vastly improved.

It was clear from the beginning that soil management and building the soil were going to be critical for us. With management-intensive grazing, we want to have some of that forage trampled because it actually goes back into the soil as organic matter.

Farmscaping

Farmscaping is the integration of other types of plants (such as trees, shrubs, forbs, and grasses) into your farming system to increase the population of beneficial insects or for other natural resource conservation benefits. Buffer strips and hedgerows are two examples of farmscaping.

Buffer strips are sections of a farm field or border that are maintained in permanent vegetation in order to enhance soil, water, and air quality. Buffer strips trap sediment, hold soil in place, and reduce surface runoff. The resulting improvements in soil water infiltration provide multiple benefits for farmers and ranchers. Types of buffer strips include grassed waterways, contour strips, vegetative barriers, and filter strips. Windbreaks may also be considered a type of buffer strip.

Hedgerows (or other field border plantings) are any mix of trees, shrubs, or grasses that exist along the edges of farm fields. They may consist of plants that have established themselves naturally or that have been deliberately planted for this purpose. Either way, they provide a number of benefits for farms and ranches, including erosion control, improved water quality, enhanced biodiversity and wildlife, attraction of beneficial insects and pollinators, and weed control along field borders.

To establish effective hedgerows and buffer strips, you need long-term planning and management. Important steps in this process are to:

- Develop a farm plan
- Carefully evaluate the site; determine where fields, roadways, fencelines, and other infrastructure will be located
- Select the appropriate plant mix
- Prepare the soil and plant the hedgerow or buffer strip
- Have a plan for watering the plants and for weed and rodent control, especially during the establishment period

Once established, hedgerows and buffer strips usually do very well with just annual maintenance to keep weeds under control. Other types of in-field intercropping practices (strip cropping, alley cropping, and insectary plantings) are also used to encourage the introduction of and increase in beneficial insects.

PUTTING IT ALL TOGETHER

In this chapter, we have discussed how growing crops or raising livestock involves managing a whole ecosystem that includes energy flow and different cycles (water, carbon and other nutrients, and organism life cycles) that affect its productivity. The complexity of a farm ecosystem presents opportunities and challenges. How do you manage a biological system where the life cycles and processes of the many organisms that inhabit that space are interconnected, constantly in flux, and finely tuned to changes in environmental conditions? And how do you make that system profitable?

Despite this complexity, farm ecosystems can be managed in such a way that increases productivity and profitability. The basic approach described in this chapter is rooted in the concept of trying to work with nature as much as possible (as opposed to trying to control it).

Farmers Share Their Experience

RECORD KEEPING FOR PRODUCTION

ONE OF YOUR MOST IMPORTANT TASKS as a farmer is keeping good records. As you read these farmers' comments, note how keeping good records helps them become better farm ecosystem managers. Think also about how essential record keeping is when implementing the key practices covered in this section of the chapter. Good record keeping will help you do the following: stay organized and get everything done on schedule, remember what you have done and track your progress from year to year, know what practices are working on the farm, identify problem areas and the effects of microclimate, and track production costs and improve your overall profitability.

JOSH VOLK, SLOW HAND FARM (profile on page 46)

Over the years that I've been farming, I've developed what I think is a pretty sophisticated crop-planning method. My crop plans develop to-do lists for me for the entire year and then those lists are also made into record-keeping sheets. So essentially as I'm checking something off, I'm also writing a record of what I did and when I did it.

At least once a week, I update all of my to-do lists and keep the records from the previous week. I'll record anything that was different and what day that planting actually happened. I then have record sheets that I can use to plan subsequent seasons. I also have to-do lists for the greenhouse, and I keep track of what goes in the share every week.

SLOW HAND FARM

I use a pretty simple, straightforward set of maps of everything that's on the farm and where it is. These maps help me keep track of when crops go into particular fields and when they come out, and when significant things like cultivations happen to particular crops. I use those maps the following year to develop my crop rotations and plan for planting and harvesting dates.

One of the tools that I use is a smartphone. I keep records in the calendar on the phone, which works well because I can search for a term like "garlic harvest" and figure out when that happened in previous years. That's been a really good record-keeping tool for me. I also take pictures around the farm with my smartphone and use the time stamp on the photos to keep track of when things happened.

CARLA GREEN AND MIKE POLEN, SWEET HOME FARMS (profile on page 216)

Mike: We have tried various things to keep track of more detailed information on the animals, like birth weights and dates and individual animal identification numbers. We ended up with a software program called Ranch Manager. The tag numbers of all the cattle, goats, and sheep are logged in as they're born.

We keep notebooks in each barn and truck, so that we can write things down as we do them. Then later we enter that information into the management software program, so that we can make decisions like when to breed and whether to cull certain animals, depending on their birthing or illness history.

Carla: You have to keep records in order to be able to assess how and what you're doing. This is not an easy business to be in to be profitable, so you have to look at your costs and your production across the board. And the only way to know these things, and whether your business is profitable, is to keep track of the numbers.

BETH HOINACKI, GOODFOOT FARM (profile on page 44)

I may go a little overboard in my record keeping, but I have found that good records make me a better farmer. It doesn't matter how you keep your records, or what they look like. The important thing is that you keep them. I carry a little pocket notebook and a pencil or a pen everywhere I go.

Every morning, I go out in the field and that notebook does two things for me. One, it tells me what I need to do that day. And two, it allows me to write down what I need to do in the upcoming days. When I'm out in the field, things I need to do will come to mind, and if I waited until I got back to the house to write them down, I'd forget them. That notebook is my key to working smarter, not harder, basically.

Back at my desk in the house, I have larger books. I have an 8½ by 11-inch notebook that is my field activity book. In it I write down everything that I did on a given day. I also have logbooks. I have an irrigation log in which I note the date, where I watered, and for how long I watered; a grazing log, where I note the date and which animals I moved out of and into which areas; and a picking log to track harvest dates and weights.

I keep notes of how I would have done some things differently and what I need to watch out for next year — things that give me a heads-up in the future. So that notebook captures not just what I do and when I do it but also things that are going to be part of the planning process for the following year.

GOODFOOT FARM

SWEET HOME FARMS
BUILDING A SECOND CAREER

Sweet Home Farms was an 82-acre livestock ranch in the foothills of the Cascade Mountains in Oregon owned and operated by Carla Green and Mike Polen from 2006 to 2012.

Prior to starting the farm business, Carla and Mike were looking for a change from the daily grind of their office jobs and took inspiration from Michael Pollan's book *The Omnivore's Dilemma* and his description of Joel Salatin's "stacked system." "It was pretty interesting to us — this notion that you could raise different kinds of animals on the same pasture, and that they would be synergistic and help improve the land," remarks Carla. "And on top of that, there was evidence that this type of system could be more profitable for a small farmer."

Carla and Mike did their homework, contacted their local Oregon State University Extension advisor, and started visiting different farms to learn more. They also began looking into the finances and figuring out the economics of a farm of this type. "We knew we wanted to raise pastured poultry," says Carla, "so it was key that we be situated near a local poultry-processing facility." That factor eventually led them to the home property near Sweet Home, Oregon.

Over the first few years of the business, Carla and Mike slowly increased their land base,

leasing additional properties that would allow them to produce more product. They also worked on developing and improving their production system. At its peak, Carla and Mike managed about 500 acres total (82 acres that they owned, plus another 400 acres of leased land) on which they produced beef, lamb, goat, pork, and chicken. Meat products from the farm were frozen and marketed direct to customers at the farm and online. They also ran a meat CSA from the farm.

Carla and Mike developed a "stacked system" where different types of animals and meat products were produced on the same land. The system featured rotational grazing with electric fencing, a mobile chicken house, and dogs for herding and guarding the livestock. Says Carla, "Our goal and guiding principle through this time was to have a sustainable farm with humanely raised animals that produce high-quality meat."

Their commitment to sustainability embraced not only the farm business but their personal lives. After several business challenges, Carla and Mike sold the farm and moved to Gaston, Oregon. They now have a small farm there where they raise working dogs, sheep, and a few goats.

"Our goal and guiding principle through this time was to have a sustainable farm with humanely raised animals that produce high-quality meat." — CARLA GREEN

RAINSHADOW ORGANICS
COMING HOME TO THE FARM

Rainshadow Organics
OREGON

Sarahlee Lawrence manages Rainshadow Organics, a 130-acre organic farm in Terrebonne, Oregon, with her husband Ashanti Samuels. Sarahlee grew up on the farm, then attended Whitman College in Walla Walla, Washington, where she studied environmental science.

After graduating from college, Sarahlee left the area and spent 10 years rafting rivers all over the world before returning to the farm. Historically, the farm, as managed by her parents, was a Longhorn cattle and hay ranch. But that was not Sarahlee's vision. As Sarahlee puts it, "I was floating on several hundred miles of all of the tributaries of the Colorado watershed and reading Michael Pollan, and really coming around to the idea of farming, and farming food."

In returning to the farm, Sarahlee realized that it was her job to come up with her own business. She started with two acres that had access to water for irrigation (a necessity to farm in the high desert) but had never really been farmed. That allowed her to grow vegetables organically right

from the start. She planted cover crops to help build the soil and began exploring other sustainable production practices. The farm has grown over the years, and Sarahlee now works about 27 acres — half in row crops (a variety of vegetable crops, potatoes, winter squash, and dried beans) and half in cover crops, wheat, and other grains.

Another major component of the business is meat production (pork, beef, chicken, and turkey). All products from the farm are sold locally in central Oregon at farmers' markets, through the Rainshadow CSA, and at the farm store.

The farm is located in a high desert environment, in the rain shadow of the Cascade Mountains. Even in this challenging climate — only seven to nine inches of rainfall per year — Sarahlee has achieved remarkable success. She talks about the guiding factors that have helped her achieve her goals: "I have a strong sense of place and being connected to the ground that I'm from. I'm rooted here in a way that I can't really explain. So that's a big part of why I'm here. And also my love of my community, and wanting to be able to provide that community with healthy food. Lastly, I have really pulled way into what I have control over and how I can make this world a better place in my daily actions. And that is why I do what I do."

KEEP IT

CHAPTER SIX

ENTREPRENEURSHIP, FAMILY BUSINESS DYNAMICS, AND MANAGING RISK

How to keep your farm going for the long haul is the theme of this final chapter. There are challenges to operating any small business, and farms are no exception. Some might say that farm businesses are even *more* challenging, given the number of variables that you have to deal with in producing, marketing, and selling your product. Many of the challenges related to maintaining a farm business are reflected in the following learning outcomes.

BY THE END OF THIS CHAPTER, YOU WILL:

- Understand your role as an entrepreneur and be able to assess new business opportunities.
- Have an idea of how you would like your business to grow, and start to develop a business growth plan for your farm.
- Understand your options for how to legally structure your farm business.
- Describe the major roles in a family business that contribute to your satisfaction and the long-term success of the business.
- Understand the differences between a succession plan and an estate plan, and evaluate your own farm business from that perspective.
- Know the basic regulations and licensing requirements for growing and selling food.
- Understand the importance of risk management in protecting and strengthening your farm business, and identify steps you will take to address risk on your farm.
- Have a realistic view of farming and a better understanding of both the challenges and the rewards of that vocation.

These outcomes are addressed in six major sections: Entrepreneurship, Business Structures, Family Business Dynamics, Licenses and Regulations, Risk Management, and Reality Check. Before proceeding, however, take a moment to review some of the major highlights from previous chapters:

- In chapter 1, "Dream It," you explored your vision of what it means to run a small farm business. You developed a mission statement and goals, and you evaluated the resources you have to work with.
- In chapter 2, "Do It," you learned how to put those resources to work, looking at the practical aspects of equipment and infrastructure, irrigation, labor management, and farm energy.
- Chapter 3, "Sell It," was based on the premise that before deciding what you want to grow or raise on your farm, you need to know if (and how) you can sell it. You learned about key marketing concepts and identified the marketing strategies that are most compatible with your own farm business.
- In chapter 4, "Manage It," you worked through the key steps in managing the financial health of your farm business.
- And in chapter 5, "Grow It," we covered the work of managing your land and resources to produce crops and livestock.

In "Keep It," we integrate these diverse elements with an emphasis on achieving long-term success for your farm business.

ENTREPRENEURSHIP

It is an exciting time to be starting a small farm business. Small farms are an important sector of the farm economy. According to the National Institute for Food and Agriculture, small farms and ranches are vital to "the competitiveness and sustainability of U.S. rural and farm economies." Small farms are also a major force in recent trends toward locally produced food. High-value products are being developed and marketed through community-based approaches, where farmers are working with other farmers, retailers, chefs, and their customers.

Creating and managing a successful farm business requires an entrepreneurial mind-set. In the simplest terms, this means being nimble and recognizing and acting on new opportunities (even in uncertain conditions) and consistently questioning every aspect of your business's operations. Some people have a natural ability for this kind of thinking, but it is also a skill you can develop with practice. Two important concepts that can help you develop an entrepreneurial mind-set are cognitive adaptability and decision frameworks.

Cognitive adaptability is the ability to sense and process changes in the environment, then be flexible in responding to those changes. To increase cognitive adaptability requires that you think about the way you think. Research has shown that entrepreneurs with higher cognitive adaptability are better able to adapt to new situations, be creative, and communicate their reasoning behind a particular response. In terms of your farm business, this means learning to read and evaluate the market, interpret that information in light of your knowledge and experience, and develop a plan of action to take advantage of new opportunities.

Decision frameworks are the mental constructs and strategies that help us organize prior knowledge and experience to make sense of what is happening and make an informed decision. A decision framework can be as simple as drawing up a list of pros and cons. Or it may be a more involved process, like the SWOT analysis you worked on in chapter 1, "Dream It" (page 42). Some approaches to farm or ranch management might also be thought of as decision frameworks. Clearly, whole farm management is a decision-making framework.

Identifying and Assessing an Opportunity

How do you identify and evaluate a new farm business opportunity? Good business opportunities do not usually drop ready-made into your lap. In most cases, you have to create them. That means being alert to all the possibilities and having a system or method to evaluate whether an opportunity is going to work for you.

Ideas and inspiration for the farm business often come from people you interact with regularly: other farmers, customers, family or friends, retailers, wholesalers, or processors. People often say, "If only there was a product that would . . ." Listening for these kinds of statements can spark your thinking about new business opportunities.

Once you have identified an opportunity — a new product or value-added component, new ways of distributing and creating convenience, or new markets — then you need to critically evaluate it. There are various methods for evaluating potential business opportunities, but the basic process should address the following key questions:

- What is the period of time for this opportunity? Is there just a brief window of opportunity, or is the new venture viable over a longer period?
- Who is your target customer? What is the potential market size?
- What are the potential costs and returns of the new opportunity?

- What capital or other resources would be required?
- What unique qualities of this new opportunity (or product) would give you an advantage compared to your competitors?
- How does the opportunity fit your business? Is it consistent with your vision, mission, and goals?
- How passionate are you about this opportunity? Is your commitment toward the new opportunity strong enough for you to make the sacrifices that may be necessary to reach your goals? (Although many entrepreneurs think that their energy and motivation will increase as the venture is developed, typically it does not work out that way. It's best if you have a high level of motivation to start with.)
- Do you have the skills and capacity within your farm team to achieve success?

Assessing the Opportunity

After you have identified an opportunity that you want to pursue, the next step is to assess whether or not to act on the opportunity. Below is a summary of the four segments of an assessment.

Product description and context. Describe your product/service idea, analyze the competition, and determine the uniqueness of your idea in terms of selling it. Is there both market need and demand? Will people desire your product/service over what is already on the market?

Market. Consider the market characteristics, trends, and growth. What is the need for your product in the market or the problem it will solve? Who are the potential customers? Why will they want your product? What can you reasonably charge?

Your business team. Describe your farm team in terms of skills, experience, and expertise related to the product/service. Does the product/service excite you and your team? Does it fit with your prior experience? What additional skills are needed to make the product/service successful?

Timeline. Indicate the steps you need to take to successfully bring your idea to reality. Put the steps in sequential order. What time and money are needed for each step and what is the total time and money needed?

Determining the Required Resources

What resources will you need to develop your business idea and bring it to fruition? Is capital the most important resource? Or is it labor? How can you systematically determine the areas where you are currently prepared and those areas that need more attention? The process of developing a business plan is the best way to find the answers to these questions.

Even without a formal business plan, you can begin the process of determining what resources are needed for developing a business opportunity by evaluating the following:

- **Tangible resources:** Physical resources on the farm, such as capital, land, buildings, plants, equipment, and supplies
- **Intangible resources:** Resources that you cannot see or touch, such as your reputation, brand, and intellectual property (patents, copyrights, trademarks, trade secrets)
- **Capabilities:** Human resources and skills represented in your farm team, such as production expertise, marketing expertise, business and financial management skills, customer service, ability to hire, motivate, and retain employees

You have done some of this type of evaluation already in the farm plan worksheets you worked on in chapter 1, "Dream It" and in chapter 2, "Do It." In particular, note the following components of your farm plan:

- Skills assessment
- Land resource inventory
- Infrastructure and equipment assessment
- SWOT analysis

Once you have a clear picture of your existing resources, then you are ready to develop a list of additional resources needed to fully implement your business idea. That list can also include tangible resources, intangible resources, and capabilities. This process is closely related to developing a budget, which you learned about in chapter 4, "Manage It." Again, take care to not underestimate the amount and variety of resources needed, including the amount of labor. In addition, any resources that are vital should be differentiated from those that are just helpful. Also assess the risks that are associated with inappropriate or insufficient resources.

It is important at this stage to think about how you will acquire new resources you need to fully develop your business idea. Acquiring tangible resources such as land, capital, or equipment may involve some type of financing. Refer to chapter 4, "Manage It," for a more complete discussion of financing your business. Developing intangible resources might involve networking with other farmers or building social capital (covered in chapter 1, "Dream It"). To increase your farm team's capabilities, some workers may need to go back to school or receive other types of training and professional education.

Growing the Business

Growing your business brings both benefits and challenges. As the business changes, pay particular attention to the following:

Cash flow. Regularly prepare monthly (or even weekly) cash flow statements in which you compare actual and budgeted amounts. Make adjustments to your budget when necessary to more accurately reflect actual business conditions. We covered cash flow statements in chapter 4, "Manage It."

Inventory. Managing inventory begins with careful planning before the start of the growing season. Your goal is to match production levels with potential sales. This is particularly true for perishable commodities. For other types of products that have a storage or shelf life, some amount of inventory is appropriate. Plan carefully here, since you must determine the amount of product you need and over what time frame. Too much inventory can be a drain on your cash flow since the money you invested in producing your product is in storage, waiting to be sold. Yet too little inventory leads to stock-outs and lost sales.

Fixed assets. Fixed assets can be expensive and have regular costs associated with them. Equipment will need servicing and insurance, and it will depreciate over time. As we

discussed in chapter 2, "Do It," consider renting equipment or hiring out for certain services. This approach can lower your costs and may provide a tax advantage since the expense is a tax deduction.

Costs and profits. Monitoring your costs and tracking profit margins from year to year are important tasks in managing your farm business. By keeping up with these analyses and with financial statements, you can better evaluate if new business ventures are paying off, or if you need to make adjustments. For more information on cost control and financial statements, revisit chapter 4, "Manage It."

Taxes. Remember that as your business grows, you will have a higher tax liability and perhaps be in a higher tax bracket. You will need to withhold and pay more tax, such as federal, state, Medicare, and social security.

Human resources. As your business grows, you may need to bring on additional employees and seasonal help, which will require you to spend more time and resources on labor management. To manage this area of the business, you will have to find and hire employees; determine whether employees should be full-time, part-time, or seasonal; create and implement a fair employee evaluation process; and maintain (or strengthen) the workplace culture of your farm business (such as camaraderie among employees, positive feedback, training and education, and opportunities for sharing in the business's success). Revisit chapter 2, "Do It," for more information on hiring and managing workers.

Your time. Finally, as your business grows, you may feel more pressed for time. Managing your schedule and commitments will take effort and determination, but with improved time management you may enjoy increased productivity, increased job satisfaction, improved personal relationships, reduced time anxiety and tension, and better health. Improving time management often centers around the following principles:

- **Intention/motivation:** Recognize your need to change attitudes and habits regarding your time allocation.
- **Effectiveness:** Focus on your most important tasks, not necessarily the most urgent ones.
- **Analysis:** Understand how you are currently spending your time and where it is being spent inefficiently.
- **Teamwork:** Working as a team is powerful; work on building team efficiencies and utilize the skills of each member.
- **Prioritized planning:** Categorize your tasks by their importance, then allocate your time accordingly.
- **Reanalysis:** Periodically review your time management process. How are you doing?

BUSINESS STRUCTURES

An important aspect of securing the future of your farm business is to establish the right form of business ownership. How you structure your business could affect your taxes, your ability to make timely decisions, your liability, and your farm succession plan. When deciding how to structure your business, consider the following:

- The nature of your business
- Your vision regarding size and growth
- Your desired level of formal structure and control
- Business and personal liability to lawsuits
- Tax implications of the various structures
- Expected profit or loss
- How you expect to reinvest your earnings
- Personal cash needs

An excellent resource for exploring these issues is "Start and Grow Your Business" from the U.S. Small Business Administration's Business Guide (see Resources).

Let's look now at some of the most common forms of farm business ownership: sole proprietorships, general and limited partnerships, limited liability companies, and limited liability partnerships.

Sole Proprietorships

Most small businesses begin as sole proprietorships. They are usually owned by the individual who manages the business day to day. As a sole proprietor, you own all of the business assets as well as all of the profits created by the business. You will have great freedom of action, and your business is taxed at your individual tax rate and not a corporate tax rate.

However, you also assume complete individual responsibility for any of the business's debts or liabilities. In the eyes of the public and the law, you and the business are one and the same entity. You may have difficulty obtaining the capital needed to manage and improve your business, and you may not have the skill set required to maximize your business's chances of success.

SOLE PROPRIETORSHIPS	
ADVANTAGES	**DISADVANTAGES**
Privacy	Limited capital
Unique tax advantages	Difficulty in obtaining credit
Owner doesn't have to share profits	Inadequate management and employee skill
Relative freedom of action and control	Unlimited liability for firm's debts
Easiest and simplest form to organize, operate, and dissolve	Limited business life because business and owner are legally the same

Some federal tax forms needed for proprietorships include:

- Form 1040: Individual income tax return
- Schedule C: Profit and loss from business
- Schedule SE: Self-employment tax
- Form 1040 EZ: Estimated tax for individuals
- Form 4562: Depreciation and amortization
- Form 8829: Expenses for business use of your home

For more information, consult the Business Guide from the U.S. Small Business Administration (see Resources).

Partnerships

In a general partnership, two or more people share the ownership of a single business. As with proprietorships, the law will not distinguish between the business and you and your partners. Legally, the partners are treated as equals and are liable for the actions of other partners. To establish a partnership, you should have a legal agreement that stipulates:

- How business decisions will be made
- How any disputes will be resolved
- How profits will be shared
- How you will allow future partners to enter
- How you will be able to buy out a partner
- How the partnership can be dissolved if needed

Some partnerships opt for an equal partnership in which ownership is split 50-50. Yet in equal partnerships, no one has a final say and decision-making can quickly bog down, so a 51-49 split may make more sense. The Uniform Partnership Act (UPA) governs how partnerships operate in the absence of other expressed agreements. More details about the Uniform Partnership Act can be found at the Uniform Law Commission website (search for the Partnership Act of 1997).

PARTNERSHIPS

ADVANTAGES	DISADVANTAGES
Easy to form	Limited business life
Division of labor and management responsibility	Unlimited liability for debts of the firm
Can use ideas and plans of more than one person	Each partner is responsible for the acts of every other partner
Specialized skills available from all partners	An impasse may develop if the partners become incompatible
Can raise more capital since good credit may be available	Death of any one of the partners terminates the partnership
Obtains financial resources from more than one person	A partner cannot obtain bonding protection against the acts of the other partner(s)

Some federal tax forms needed for partnerships include:

- Form 1065 K-1: Partner's share of income, credit, deductions
- Form 4562: Depreciation
- Form 1040: Individual income tax return
- Schedule E: Supplemental income and loss
- Schedule SE: Self-employment tax
- Form 1040 ES: Estimated tax for individuals
- A good place to find more information is the Business Guide from the U.S. Small Business Administration (see Resources).

In a limited partnership, one or more general partners conduct the business while one or more limited partners provide capital, have no management participation, and are not liable for the debts of general partners.

Limited Liability Structures

Limited liability company. Limited liability companies (LLCs) are formed to gain the tax advantages of a partnership with the limited liability of a corporation. They are now permitted in all states, with slight variations in each state. Generally, if you want to form an LLC, you will need to choose a business name, file articles of organization, and create an operating agreement. To learn more about limited liability structures, check with the office in your state that handles business registrations (usually the Secretary of State). For specific information on LLCs, visit websites of the Internal Revenue Service, Small Business Administration, and Nolo.

Limited liability partnership. A limited liability partnership is used to protect individual partners from personal liability of negligent actions by other partners or employees. Limited liability partnerships (LLPs) are not recognized in every state, and those that do sometimes limit LLPs to professional services like law or medicine. For more information on LLPs, visit the Entrepreneur and Nolo websites.

Family limited partnership. As the name implies, a family limited partnership (FLP) is where the majority of partners are related to each other as spouses, parents, grandparents, siblings, cousins, nieces, or nephews. Most family limited partnerships are family farms because of the explicit involvement of family members in the business. FLPs offer the same tax advantages as LLPs, and in some states, FLPs have additional benefits.

COMPARISON OF BUSINESS STRUCTURES

STRUCTURE	TYPICAL OWNER	LIABILITY	TAXABLE PROFITS	PAPERWORK OF ORIGIN
Proprietorship	Individual	Individual	Individual	None
General partnership	2 or more	Individual	Individual as per partnership agreement	Partnership agreement
Limited partnership	2 or more; 1 general partner	General partner	Individual as per partnership agreement	Partnership agreement
Limited liability company	2 or more	Business entity	Choice of individual, partnership, or corporation	Charter; articles of incorporation
Limited liability partnership	2 or more	Business entity	Individual as per partnership agreement	Varies by state
Family limited partnership	2 or more	Business entity	Individual as per partnership agreement	Varies by state

FAMILY BUSINESS DYNAMICS

There are many different types of farm ownership arrangements. The different business structures you learned about in the previous section reflect that variety to some degree. Yet even within those legal definitions of the business, there are diverse models of ownership. Think of the different types of partnerships that you know of: between family members, between friends, partnerships based solely on business interests — or any combination of these. As we begin this section on family business dynamics, think also about your definition of family. In using the term "family" here, we acknowledge how that concept encompasses a variety of relationships and living arrangements. *How do you define family? How closely is your family involved in your farm business?*

Think of your spouse or partner, your children, your parents, or other relatives. Even when farms are owned and managed by an individual, or when the people in the business partnership are unrelated, other family members are usually involved in some way. Perhaps they are directly involved as co-owners and comanagers, as employees, or as seasonal help. Or maybe they helped finance the farm, or they are loving and supportive family members who work off the farm to provide added income and health benefits for the family.

Renard and Chinette Turner, Vanguard Ranch

Now reflect back on chapter 1, "Dream It." Think about your vision for your farm and your quality-of-life assessment. In what ways are your feelings about family reflected in those statements? For many farmers, much of the attraction of owning an independent business is the opportunity to live every day according to your closely held values. Those values often include spending quality time with your family and being present for them.

Sometimes, because these values are so heartfelt, it can be challenging to own and manage a small family farm. Families strive for harmony and inclusion with unconditional love and support for family members. Businesses thrive on efficiency and achievement, promoting individuals based on their job performance. These can be competing interests, causing tension.

In this section, we look at best practices to keep your business and family strong and avoid potential conflicts down the road. These practices are the building blocks of good communication and sound planning that will help you realize the benefits of family farm ownership. Those benefits are well documented:

- Family-owned farms become strong supporters of the local community by hiring local workers, supporting community causes, and generating economic activity in the local area.
- Family-owned and -operated businesses have efficiencies so they can make operating changes quickly or collectively pitch in to endure an unexpected emergency.
- In family-owned businesses, leadership tenure is almost 18 years compared to just 8 years for public companies. A longer time horizon for planning and investment is an advantage in establishing trust and strong working relationships with business contacts.

Left: Beth Hoinacki and her children, Goodfoot Farm; **Middle:** Chris Jagger and Melanie Kuegler and their children, Blue Fox Farm; **Right:** Mimo Davis and Miranda Duschack's son, Urban Buds

Roles in the Family Business

To become more aware of the potential challenges of running a family business, it helps to understand the different characteristics of "the nurturing family" and "the competitive business." One way to examine these differences is to look at the three roles that family members commonly take on in a farming business.

Family role. Even if they don't work on the farm, family members are a key part of the supporting family. Families involved in the farm can include a single generation, such as husbands and wives, or siblings. Multigenerational family farms include at least two generations working together.

Farm partner role. This is when family members are interested in supporting the farm operation as a part owner or investor. Farm partners and other family members may not necessarily work in the business.

Manager/employee role. Family member managers and employees have the day-to-day responsibilities for operations of the farm business.

Many families find that clearly defining their roles and responsibilities helps to limit the conflict and frustration in family businesses. For example, the partner (shareholder) family members may want to increase the dividends and income from the farm, while those in a managerial role may want to reinvest the earned income to upgrade farm equipment. Or maybe the farm owner/manager wants to work every Sunday, but other family members want to have more time together. Analyzing conflicts based on the family members' roles in the business provides deeper understanding of each member's perspective and ultimately creates the foundation for a long-lasting family businesses. To explore how these roles might overlap, work on the Family Business Roles Activity worksheet (appendix 3-1). Copy and fill out the chart for your own farm business. Have other family members involved in the business fill one out, too, and compare answers.

FAMILY COMMUNICATIONS

More and more families are using the family meeting as a forum to develop good communication, address issues within the family, ask questions about the business, and deal with conflict. Family meetings should include all of the family members and have an agenda. They work best when held at a neutral location away from the business.

To develop a healthy work-family balance, some families make it a rule not to talk about the business after a designated time in the evening. Others make sure that the business office is in a separate location, outside the home. Other families might schedule regular vacations just so the family can reconnect and have fun together. Besides giving everyone a rest, this bonding establishes good working relationships among family members who may be running the farm as partners in the future.

FARM PARTNERS

Family farm shareholders, or owners, usually play an important part in big-picture strategic planning for the farm business. Owners are not necessarily responsible for the day-to-day operations. An important aspect of the ownership role is supporting long-term planning and continued capitalization of the business. This can be accomplished by creating a board of directors or, for smaller farms, an advisory board. Even for a small farm, with just one or two owners, long-term planning is crucial for the success of the business.

We also recommend that owners consult regularly with informal advisors — non–family members from related industries who can advise the farm and provide input. Private family business owners who use these advising teams report that the feedback makes their business stronger and helps everyone in the family improve. Regular meetings and communication with a board or advisory team can keep you on track with your goals and objectives. In turn, the business benefits from the experience and insights of your advisory team.

When there are multiple shareholders, you should establish clear instructions for what happens to ownership shares in the face of untimely death, divorce, or disability or when a family member owner wants out of the business. Consider drawing up a buy/sell agreement — a formal contract that documents the terms and conditions for selling farm ownership. This kind of agreement can include the requirement that sellers not offer farm ownership outside of the family until the family has had the opportunity to buy the available shares. It is best to develop these agreements before any crisis occurs.

Other tasks that farm partners may be responsible for include:

- Discussing proposed geographic or market expansions
- Reviewing financial statements and reports
- Developing a succession plan
- Preparing for future leadership in the business
- Updating shareholder agreements

Who would be your candidates for advising your family farm? Who might you ask to be on your formal team of legal and professional advisors?

Farmers Share Their Experience

FAMILY BUSINESS DYNAMICS

WE EXPLORE IN THIS CHAPTER some of the important issues that arise when running a family business: balancing family and business goals, defining the roles and responsibilities that family members take on in the business, family communication, and working with farm partners and shareholders. The farmers quoted here have worked through similar challenges in their own farm businesses. How does each farmer approach the concept of "roles" in the business? What strategies do they use to foster communication and understanding among family members?

CARLA GREEN AND MIKE POLEN, SWEET HOME FARMS (profile on page 216)

Mike: All the stories about the difficulties of running a family business are true. It's difficult at times. What I think works really well for us is to have common goals in the business. It's often a struggle, but we do try hard to focus on what's good for the business, and then try to keep that separate from other family dynamics.

Carla: Both of our kids work with us on the farm here. We sent them to the same ranch management class that we took so that we have a common language to talk with them about the finances and the economics of the business, as well as the water, soil, and grazing management practices. So we have that commonality.

One of the things we learned at the class was to make distinctions between what ranch management consultants call WITB and WOTB. "Working in the business" (WITB) is the day to day, on the ground, taking care of animals, and dealing with problems. "Working on the business" (WOTB) is the planning, the vision, where you're going, how are you going to make this a profitable business.

We have WITB meetings where we make decisions about production and day-to-day management. And we have WOTB meetings, which are focused on how we grow this business. And we have different roles for those different meetings.

CHRIS JAGGER AND MELANIE KUEGLER, BLUE FOX FARM (profile on page 168)

Chris: For Melanie and me, I think that we've both just kept an open line of communication with each other. And we're both very strong-willed people, too, so we've really learned a lot about patience.

Melanie: We've each taken on certain roles within the business, which we've defined for ourselves. We try not to step on each other's toes too much or micromanage each other. It's also a system of checks and balances. I mean, he comes up with some ideas and I'm like, "Dude, no." And then sometimes I may say no initially, but then we have a discussion about it and I realize that there's validity to the idea — and he goes through the same process with me.

Four of the five generations of the Kiyokawa family: Randy (3rd generation), Korbi (5th generation), Michiko (2nd generation), and Catherine (4th generation)

Father David Lawrence and daughter Saralee Lawrence, Rainshadow Organics

RANDY KIYOKAWA, KIYOKAWA FAMILY ORCHARDS (profile on page 94)

I'm extremely fortunate that my parents were gentle but then again firm. When I came back to the farm, I just worked as a laborer for the first year or two, doing everything from cutting grass to changing water, and didn't make any decisions. I was eased into the business slowly.

Over the next two or three years, I slowly made more and more decisions. And by probably the fifth year, I was allowed to make the lion's share of decisions, for better or for worse. Mom and Dad have been great. They're very conservative. I've learned to give them time to absorb my new ideas, then come back to them later with reasons why they would be good for our family operation.

SARAHLEE LAWRENCE, RAINSHADOW ORGANICS

(profile on page 218)

Coming back to the farm, my job was to come up with my own business. My dad and I help each other but we don't overlap at all on the farm. And it's wonderful to have multiple people here that are capable. I find that it's really wonderful for people to have their own work and their own jobs. My husband, Ashanti, is responsible for the animals, and he makes his own decisions. I try really hard not to worry about the animals and just let him do his thing. On occasion we cover for each other and we don't get much sleep, but otherwise we're running two separate businesses that need their own attention and total focus.

LISA MISKELLY AND ANTON SHANNON, GOOD WORK FARM

(profile on page 92)

Anton: Lisa and I took a holistic management course, and that approach really speaks to our quality of life. I think we just try to ask ourselves, Are we farming the way we want to be farming? Are we living the way we want to be living? If not, what do we want?

For us, we do our best to live fairly modestly. Time with the horses is important. We both enjoy working with the horses and having time to enjoy being on our farm, and not being burdened with too many financial worries. That's one of the reasons Lisa got an off-farm job — so that we would have a little breathing room. We try to consider where the farm fits into the complete picture of our lives, and then make adjustments to the farm. We try to do our best to not let the farm lead us.

Lisa: Which sounds nice on paper. I'm not sure that the reality is anything very close to that.

Anton: But that is the grand vision that we sometimes get lost in trying to follow.

A lot of people seem to struggle with farming with one's partner. It can be really hard on relationships, which can impact how much you're able to do and accomplish. We've certainly seen other farming couples who have really struggled in their relationships because they work together all the time.

Anton Shannon and Lisa Miskelly, Good Work Farm

Miranda, August, and Mimo, Urban Buds

Lisa: I would say that was part of the motivation for me wanting to work off the farm — to find that balance. Spending 100 percent of your time with your partner, handling the stress of production and finances, and having both of your livelihoods tied to the same business is just a lot for a relationship to handle.

And there's certainly a lot of good that comes out of working so closely together. Now that I'm not working with Anton all the time, we definitely miss each other and miss farming together. But we acknowledge that difficulties arise out of that.

Anton: I think a lot of couples who go into farming together come either from a nonfarming background or they've both worked on farms together and then want to run their own farm. I think it's important, if you're a beginning farmer, to realize that you'll be running a business with this person, which is different than just weeding and transplanting with that person. There's a difference between running a business with someone and just doing the farmwork, and I think it's important to go into that business relationship aware of that difference and potential challenge.

Lisa: And this gets to the question of how we make decisions. I feel like we haven't really mastered having separate spheres of influence, so that we're both able to make decisions without consulting the other one. We started out consulting each other about every single thing that we did, whether it was a $5 purchase or a $500 purchase. We would hem and haw over it, discussing the pros and cons and looking at the options.

We decided we needed to have a dollar amount — like maybe $200 — so that for any purchases under that amount, we don't have to ask each other's opinion. For those decisions, we feel like we have autonomy and control, and that the other person will trust and value our decision. We're still working through that process, though now that I have an off-farm job, Anton has had to make more decisions because he's not going to call me in the middle of the day.

I think it's crucial to have good communication with your partner, be that a life partner or business partner. But it's also important to trust that person, and to trust that they're making good decisions with whatever part of the farm you guys have agreed that they should run. That can be hard in the beginning, when so much is up in the air — whether you have tenuous land security or your markets aren't as secure as you'd like or you're dealing with the infrastructure that you could afford, not the infrastructure that really fits the way you're farming.

So it's a challenge that I think is inherent. You can't worry about every aspect of the farm. You can't manage it all. It takes a team to get it done.

MIMO DAVIS, URBAN BUDS: CITY GROWN FLOWERS (profile on page 129)

When Miranda said, "Hey, let's farm together," I was like, "Look, you got to know somebody really, really well before you make that commitment to farm with them." Marrying them is one thing. Farming with them is a whole different level.

Succession and Estate Planning

As you look ahead 10 or 20 years down the road, what do you envision for your farm? As you get older and approach retirement age, how do you see yourself involved in the farm business? What can you do now to ensure that the business remains viable into the future? What will happen to the farm if you should die or become incapacitated? These are critical questions to consider, and they lead into our discussion of succession and estate planning. Succession plans and estate plans differ from one another in several ways:

A **succession plan** focuses on how to turn over ownership and leadership of the farm business to the next generation of entrepreneurs. Succession plans are broader than estate plans because they address the issues of retirement and management transition, as well as ownership transition, and are activated while the current owner is alive.

An **estate plan** on the other hand provides specific instructions on what is to happen with the farm business, and other assets you own, upon your death. Estate plans involve the execution of legal instruments such as wills, trusts, insurance policies, buy-sell agreements, and/or powers of attorneys that are needed to carry out the succession plan.

Developing a succession plan is a lengthy process. Count on it taking 10 to 15 years from the time you begin initial conversations with your family and business partners until you finalize your plan for turning over the farm to the next generation. Those initial conversations are often focused on developing a shared vision for the family business going forward. That vision becomes the basis for the plan and agreements.

In this chapter, we are simply introducing the basic concept of a succession plan. For a comprehensive succession planning tool, visit AgTransitions from the University of Minnesota (see Resources). This online service provides an outline for succession plan content and allows the plan author to share access with family members and professional advisors. Each chapter has in-depth worksheets and articles to help farm owners get started. After you enter the information in each section, you can print it out as a single document.

Although we do not cover the specific steps involved in creating a succession plan in this book, there are some key points to highlight:

- Strive for clear, open, and honest communication throughout the process. Secrecy about the decision leads to misunderstandings among next-generation family members, which can interfere with their ability to work together and can even lead to legal actions among siblings.
- Work together with your heirs to create the succession plan. This will keep them informed and strengthen their feeling of ownership in the process. Also, if they are going to be involved in managing the farm business in the future, this is an important way to prepare them for that role. The complexity of a farming operation means that future owners need plenty of time to learn the business and understand their role in making the business profitable.
- Family business founders and farmers may find it very difficult to manage the transition of ownership and relinquish control of the farm. This transition is easier when the farm operator has outside interests that he or she can pursue during retirement.

As noted above, estate plans consist of the legal instruments that provide instructions on what is to happen to your assets, *including farm ownership*, upon your death. Ideally, estate planning and succession planning occur simultaneously: one informs the other. Your vision for how the farm should be passed along (the succession plan) will serve as a road map for the estate planning process and will direct business and tax professionals in drafting the required documents. You will select legal tools to prepare the estate plan that in the end will transfer an individual's assets.

But even if your succession plan is not finalized or is still in process (especially since it can take such a long period to develop) *make sure you have your wills and trusts in place*. Because the asset structures of most family farms include high-value land-based assets and lower cash flows, a lack of estate planning can mean tremendous hardship for your survivors and potentially the loss of the farm to pay large tax bills.

Experts suggest that businesses try running a "fire drill" (a hypothetical emergency situation) so that the family can practice responding to a crisis. Particularly when multiple family members are relying on the income from the farm business, it is important to have contingency plans in place.

- Is the farm prepared for unexpected events?
- What if you or your partner were to die or become incapacitated?
- What key pieces of information would be needed to run your operation in the short term if the owner were unable to perform?

Imagining and responding to a crisis or sudden loss will help you identify what you need to do to be better prepared. Include even simple things like locating keys, important documents, and contact lists. By increasing your level of preparedness, everyone involved in the business will have peace of mind that they can handle any situation that might come their way.

Do you have all the necessary estate documents completed for your farm? Do your heirs know where they can find those documents and who is your executor? If not, make it a priority to address these responsibilities today.

Farmers Share Their Experience

PLANNING FOR THE FUTURE

SUCCESSION AND ESTATE PLANNING provide a way of looking forward and thinking about the future of the farm business. Although this type of planning is complex, it is crucial to your long-term success. Here, Randy Kiyokawa and Sarahlee Lawrence share their experiences with succession and estate planning. As you read their comments, think about the different motivations behind each farmer's future planning efforts. What are some of the ways that these farmers address the uncertainty and unpredictability of life?

RANDY KIYOKAWA, KIYOKAWA FAMILY ORCHARDS (profile on page 94)

There are still three generations on this farm. My mother's still alive, I'm here, and I have three kids. So there is a lot of planning, especially for estate taxes. That's our main concern, and it could really take us for a loop if we weren't prepared.

Each year we meet with our attorney to make sure that we've got the best plan in place for somebody to come back to the farm. I have three great kids, and I think they all would like to come back.

Would I recommend that they come back to the farm? I would like them to do stuff on their own, then see how it works out later. I don't have an objection to them coming back, but sometimes when I'm working, I think, "Boy, I wouldn't wish this on my kids."

SARAHLEE LAWRENCE, RAINSHADOW ORGANICS (profile on page 218)

I'm trying to figure out how to make sure that the land that I'm investing in and spending a lot of time on is protected for myself and that I don't lose it. I'm working with my parents on succession planning and trying to figure out a legal path forward, not just going on faith, because people can change their mind. People get divorced. People die. People change. So it's really important to have a little bit of legal backup. And sometimes it's hard to broach those subjects, but anybody who is smart will.

Developing the Next Generation

Growing up in a family farm business creates opportunities and obligations for both parents and children. Raising kids to become happy and productive members of society is a challenge for every parent. In a farm family, there are particular issues that come into play.

From the manager's perspective, it is important to first determine the job skills that are needed on the farm before inviting family members to join the business. Children who have grown up in family businesses feel more confident about their skills and contributions to the business if they have had the chance to work on these skills outside of the family business and establish their own record of work performance and promotion.

As an owner/shareholder of the business, developing the next generation of farm owners should be seen as a long-term commitment. It takes years to prepare for the complexities of managing a business. As children get older and reach the age where they can take on more responsibility, a formal plan can help them identify what skills are needed, assess their own strengths, and provide direction for training and education. Children from family businesses often use nonfamily mentors or business coaches to develop their full leadership capacity.

It is important to recognize that in some family situations, parents may have doubts about their child's ability to take on the farm business. Or if there are siblings, parents may feel that one child should get a larger (fair) share in the farm business either because that child has already shown greater commitment and given more time to the business or because the child has shown a greater aptitude for, and interest in, the work. In that case, should the parents feel compelled to pass on "equal shares" in the business, or is it better to convey "fair shares"?

Answer that question for yourself, discuss it with your partner, then view the video *Fair vs. Equal Treatment of Farm Family Heirs* from the University of Vermont (see Resources), which describes the risks of giving children equal shares of the farm when fair shares might ensure a stronger future for the farm business.

What if there are no children to take over the farm? What are your options then? As part of your estate planning process, you may want to consult with one of the organizations that link up older farmers with younger farmers who are looking for land. The Farmland Information Center maintains a national list of Farm Link Programs (see Resources).

As noted previously, a major goal of the family is to provide a supportive and respectful environment where children understand the values that drive the farm business and the legacy that the founders want to leave. Creating that kind of environment requires good communication and the ability to be open and honest about each family member's desires and concerns. Parents in family farm businesses must also pay close attention to the relationship between siblings. Sibling trust is a critical factor in developing the close working relationships that are necessary for a successful business — as partner owners, as employees, or as family members.

LICENSES AND REGULATIONS

Another important aspect of maintaining a farm business is making sure you comply with the regulations that pertain to the production and sale of your products. Some regulations may require you to obtain a permit or license for certain activities. This section provides an overview of some of the most common regulations and licenses that apply to farm businesses. Be aware that regulations (and the kind of agency in charge) can vary from state to state. We focus on Oregon here to give you an idea of the types of issues and requirements that you may need to address in your own farm business. We also point to resources that provide additional information that will help you decide what steps you need to take. First, let's review some of the assumptions and principles for this section:

1. The information in this section is directed largely at small-scale farmers with an interest in direct sales: at farmers' markets, farm stands, or through a CSA; to restaurants; or to local retail grocers, like a food co-op. If you are aiming for larger, wholesale markets, additional regulations may apply that we do not cover here.
2. Regulatory compliance can seem complicated, confusing, even boring or frustrating. However, it is an important part of your job as a farmer and food seller. Ignoring regulations can lead to significant, possibly expensive problems down the road.
3. The information presented in the chapter is based on the best information available, but you should always check with the relevant regulatory agency for exact regulatory and license requirements. Also, laws and regulations can change. Part of staying legal is staying up-to-date.
4. In most states, the agencies do a good job explaining the requirements, often on their websites. However, do not hesitate to pick up the phone and call them if you need help understanding the rules. Sometimes they can even offer guidance specific to your farm.
5. We do not address every possible topic or situation related to licenses and regulations. Cover your bases. A good way to do that is by talking to other farmers doing similar things.

The production, processing, and sale of most farm-direct products are regulated by state governments, often with reference to federal requirements. State agencies implement and enforce laws, issue licenses and permits, conduct inspections, and handle violations. Examples include state departments of agriculture, environmental or natural resource agencies, and other agencies regulating food safety and public health. In Oregon for example, the Oregon Department of Agriculture (ODA) and Department of Environmental Quality (DEQ) are the two main agencies involved in regulating food production activities. You may also need to work with county-level agencies. For example, in many states county health inspectors have jurisdiction over restaurants and food service. For some foods, federal agencies are more directly involved — for example, the U.S. Food and Drug Administration is responsible for food-safety regulations, while the U.S. Department of Agriculture oversees and inspects most meat and poultry processing. In this section, we focus on regulatory issues for three different parts of your farming and food business: production, processing, and selling.

Production Regulations

As a farmer or rancher, one of your primary responsibilities is to be a good steward of the land and other natural resources. Most state departments of agriculture manage a number of regulatory programs that focus on this stewardship role and your specific production practices. We highlight the general regulatory concerns related to environmental quality, food safety, and several other key production factors.

ENVIRONMENTAL QUALITY

Protecting water quality is a major priority for environmental regulations governing agriculture. All farmers are expected to protect water quality, and various federal and state regulations cover both point- and nonpoint-source activities. Most states have agricultural water-quality management programs that address the specific needs for various subregions within the state.

The following three practices, which can have significant negative effects on water quality if not properly managed, are typically overseen by state agencies and may require permits. In all cases, consult your state's relevant agency for specific rules and permit requirements:

- **Land application of wastewater:** Except for very small producers, land application of wastewater may require a permit from your state's natural resource management agency or state department of agriculture, or other state agency. Wastewater can include dairy parlor wash water, wash water from other packing and processing operations, and field runoff that may contain pollutants like fertilizer unused by plants.
- **Composting:** Regulations addressing on-farm composting vary from state to state. Rules usually apply to all types of source material, including plant materials, crop and processing residue, manure, and animal carcasses and meat waste from on-farm processing or other mortalities. The rules are intended to manage potential risks to water quality from composting operations.
- **Confined animal feeding operation:** In most states, farmers and ranchers with animal feeding operations on their property, even small ones, and especially those using liquid manure systems, will likely need a confined animal feeding operation (CAFO) permit.

Because all farms are different, and conditions vary across each state, it may not always be clear which water-quality regulations will affect you. But remember, even when a permit is not required, the law is that farmers may not pollute ground or surface water.

PRODUCE AND FOOD SAFETY

New federal food-safety regulations for farmers, packers, processors, and handlers — mandated by the Food Safety Modernization Act (FSMA) — are now in place and being implemented by the federal Food and Drug Administration in partnership with state agencies. These regulations are designed to prevent foodborne illness related to fresh produce. Most small farms will be at least partially exempt, but they must still meet some of the requirements, including keeping certain records, and they can lose those exemptions if linked to a food-safety problem. For more on the Food Safety Modernization Act visit the U.S. Food and Drug Administration's official Food Safey Modernization Act website (see Resources).

OTHER PRODUCTION REGULATIONS TO CONSIDER

In addition to environmental quality and food-safety regulations, farmers and ranchers may need to address a variety of other regulations, depending on what they produce and the type of business. Examples include:

Pesticide use. In most states, farmers need to obtain a pesticide applicator's license to use certain pesticides on the land they manage. Usually this only pertains to restricted-use pesticides, but check with your county Cooperative Extension Office or state department of agriculture to make sure.

Air quality. Farmers may be affected by air-quality regulations that govern activities such as agricultural burning, emissions from livestock operations, or dust levels.

Land use laws. County-level zoning and state land use laws may affect whether you can farm and what farming activities can take place on your farm. Counties also vary in their approach to permitting agritourism. Check with your county or state land use department for more details.

Agricultural labor. Farmers must comply with regulations concerning worker health, well-being, and safety. The Occupational Health and Safety Administration (OSHA) has issued a number of regulations that pertain to agricultural operations. The U.S. Department of Labor has specific regulations for seasonal and migrant labor that address pay, workers' rights, and housing.

Processing Regulations

In most cases, if farmers process what they grow into a salable product, they will need a license from their state department of food safety to do that. Food-safety programs are often found within the department of agriculture or department of health. For instance, in Oregon, licenses are issued by the Oregon Department of Agriculture which defines "processing" broadly: cooking, baking, heating, drying, freezing, canning, pickling, mixing, grinding, churning, separating, cutting, and extracting.

For some foods, processing and selling may require two separate licenses. In raising meat or poultry for market, this is most often the case. For example, for farmers who want to raise pigs and sell pork at a farmers' market, the processor will need to be state licensed (and USDA inspected, unless your state has an "equal to federal" inspection program) *and* the farmers will need their own meat seller's license. For other foods, you may only need one license to process and sell the food. Examples are cheese, baked goods, eggs, and apple cider. A note on "cottage food" regulations: Many states have so-called "cottage food" laws that allow farmers and home cooks to make certain low-risk foods in their homes for direct sale to consumers, up to some revenue limit per year. The laws and licensing requirements vary state by state — see the report *Cottage Food Laws in the United States* from the Harvard Law School Food Law and Policy Clinic in the Resources section.

MEAT AND POULTRY: THE BASICS

Meat and poultry are a bit more complicated than other foods. In this section, we highlight the basic rules. You can learn more on the Niche Meat Processor Assistance Network (NMPAN) website (see Resources).

Most red meat (beef, pork, lamb, goat, domestic elk, ostrich/emu). If you want to sell meat, the livestock must be slaughtered and processed at a facility that is USDA inspected. This means that inspectors from the U.S. Department of Agriculture's Food Safety and Inspection Service are in the plant every day and examine every animal before and after slaughter. In the 27 states with inspection programs "equal to" USDA inspection, farmers can sell meats processed under state inspection. For a list of those states, see the USDA State Inspections and Cooperative Agreements (see Resources).

Ashanti Samuels of Rainshadow Organics manages the livestock enterprise for the farm.

A "custom exempt" plant can only slaughter and process livestock for the exclusive use of the owners. But this exemption can be legally used to sell a live animal or shares in it (typically quarters or halves) before slaughter directly to household consumers. It can then be slaughtered and processed by a custom-exempt, state-licensed processor. This is because those new consumers are legally the new owners of the animal before slaughter; they are not buying meat by the cut.

The "retail exemption" allows butcher shops and other retail meat sellers to do some in-house processing (cutting up larger pieces and making some value-added products like sausage and jerky) and sell the meat. They can only do this with meats slaughtered and processed under federal or state inspection. More information about these exemptions is in the *FSIS Guideline for Determining Whether a Livestock Slaughter or Processing Firm is Exempt from the Inspection Requirements of the Federal Meat Inspection Act* (see Resources).

Poultry (chickens, turkeys, ducks, geese, or guinea fowl). As with red meat, the USDA has primary jurisdiction and inspection authority over poultry processing. To be sold as human food, poultry must be slaughtered and processed in a USDA-inspected (or "equal to" state-inspected) facility.

However, federal law allows for some exemptions to this requirement, which most (but not all) states recognize to some degree. For example, in Oregon, poultry processors can slaughter and process up to 20,000 birds each year in a state-licensed facility that is not federally inspected. Exempt processors still have to follow sanitation, record-keeping, and other rules.

PICKLES AND PRESERVES

Most fruit and vegetable processors have to be licensed and comply with extensive food-safety regulations. Farmers who want to sell value-added foods made of fruits and vegetables to stores, restaurants, or institutions must be a licensed food processor or co-pack with a licensed processor.

As noted above, Oregon and some other states allow another option to farmers who only want to sell these kinds of products direct (and who can keep it under $20,000 in annual sales): they may make and sell certain "value-added, shelf-stable products" for direct sale to consumers without being licensed as food processors. The list includes syrups, jams, preserves, jellies, canned fruit, pickles, chutneys, relishes, sauerkraut, and some salsas. As an example, details of this exemption in Oregon, including the definition of "acidic" and labeling rules, are explained in *Farm Direct Marketing, Producer-Processed Products* from the Oregon Department of Agriculture (see Resources).

RAW MILK

Thirty-one states allow the intrastate (not over state lines) sale of raw milk under certain conditions. The comprehensive State Milk Laws, published by the National Conference of State Legislatures (see Resources) contains information about the relevant federal regulations.

For instance, in Oregon, producers may have no more than two producing dairy cows (and no more than three cows total), nine producing sheep, or nine producing goats on the premises where the milk is produced. A license is not required. The farmer may advertise the product, but sales can only occur at the farm. Each state regulates the sale of raw milk in different ways, so do not sell raw milk before checking the rules in your state.

Note: The fact that some states allow residents to sell certain volumes of raw milk does not eliminate their liability if someone gets sick from drinking that milk. Study the Risk Management section of this chapter and know the potential implications of your decisions before you implement them.

...

You grew it and processed it. What about selling it? Can there possibly be any more licenses to get or rules to follow?

Selling Regulations

Farm businesses are subject to the same laws and regulations for intra- and interstate commerce that apply to all businesses. Beyond those basic rules, farmers do not typically need a license to sell fresh, unprocessed fruits and vegetables. If you are processing, most processing licenses also allow farmers to sell the finished product. For example, if you have a cottage food processing license to make pickles or jam, the license also allows you to sell those products.

Meat and poultry have more stringent regulations. To sell meat and poultry, farmers may need a meat seller's license. Specific rules and exemptions vary from state to state, so check with the agency in charge. For example, Oregon requires such a license to sell meat and poultry, except for "on the hoof" or "locker meat" sales. And if you are already a licensed processor, you do not need an additional meat seller's license to sell meat.

Other products may require specific licenses. For example, if you are selling plants or flowers, you most likely need a nursery license. Again, requirements and regulations vary by state. In Oregon, you need a nursery license from the ODA Plant Program if you sell more than $250 worth of plants per year; you do not need a license to sell cut flowers.

For produce and other bulk products, if you want to price and sell your product by weight (for example, $5 per pound for tomatoes) you will need a licensed scale. In most states, scales are licensed through the department of weights and measures, or some equivalent. If you price your product by volume or container size (for example, $5 per pint of blueberries), a licensed scale is not required. However, most diversified farms that sell direct are going to have at least a few products they sell by weight, so plan on getting a good scale and the license to go with it.

Licenses and Regulations, in a Nutshell

In wrapping up this discussion of licenses and regulations, let's revisit a few important concepts:

- If you want to grow and sell food, you need to comply with regulations around growing, processing, and selling, and you must get the relevant licenses. We understand that this part of farming is not very fun and can be frustrating sometimes. But it is an essential part of running your farm business successfully and for the long term.
- Go to the source: *always* check with the relevant regulatory agency for exact regulatory and license requirements before you get started. Feel free to ask them for help in understanding their rules.
- Laws and regulations can change. Part of staying legal is staying up-to-date.
- If we didn't address it here, that doesn't mean it doesn't exist — cover your bases. Talk to other farmers, farmers' market managers, local Extension agents, and others who can help.

KIYOKAWA FAMILY ORCHARDS

URBAN BUDS

Farmers Share Their Experience
LICENSES AND CERTIFICATIONS

LAWS AND REGULATIONS touch on almost every aspect of running a farm business. Keeping track of regulations may seem like the least enjoyable aspect of being a farmer, but it is an important part of the picture. These quotes reveal how some farmers have dealt with the complexity of the regulatory environment. As you read them, note how you might go beyond just "following the rules" and view regulations and certification as a way of distinguishing the quality of your business and products.

RANDY KIYOKAWA, KIYOKAWA FAMILY ORCHARDS (profile on page 94)

We have about 30 different scales that we certify every year. I also make sure we're food-safety certified through GAP [Good Agricultural Practices]. We have been through global GAP, too.

CARLA GREEN, SWEET HOME FARMS (profile on page 216)

We have our scales licensed and we have a meat seller's license. Those are checked, re-upped, and inspected yearly.

BETH HOINACKI, GOODFOOT FARM (profile on page 44)

I have philosophical issues with the GAP certification and the Good Agricultural Practices. I don't think it's scaled very well for the small grower, though I do think a lot of their food-safety pieces are pretty important and it is an important program.

If we were wholesaling, companies are now requiring their producers to be GAP certified. It's not necessarily a bad trend, but I think a grower has to weigh the benefits and the costs and figure out if it's a good match.

CHRIS JAGGER, BLUE FOX FARM (profile on page 168)

Now that now that we're moving toward GAP certification, we're seeing a whole new set of requirements that we don't have to do for organic certification. But it's kind of the same thing that we've seen with organic certification. Honestly, at the end of the day, it makes us better growers because we have a better idea of what the heck we're doing and our processes.

GOODFOOT FARM

BLUE FOX FARM

MIRANDA DUSCHACK, URBAN BUDS: CITY GROWN FLOWERS (profile on page 129)

There were city licensing and permit hurdles we had to maneuver. I think this piece is really important, especially if a grower is focusing on urban farming. Even in suburban and peri-urban farming, you need to make friends with your elected officials. We worked closely with the St. Louis City Business Assistance office, and we had a case worker there. Together we learned how to navigate the rules, because we were the first for-profit commercial farm in the city.

The city doesn't allow us to sell from the shop, so we have to have an off-farm retail space. We have to go to farmers' market. We don't need to have a business license. In Missouri, if your only retail spot is either on a farm or farmers' market, you don't need a business license. But before that was all figured out, we were getting notices from the city that we needed to have our business license. So it just took a lot of navigation and time.

Also, we were only the second property in the city limits to be zoned agricultural, and that really reduced our tax rate. Because we are a commercial property in a residential neighborhood, we must be issued a variance to have a business here. At first we were being taxed at a business rate, so I had to work with the city's tax assessor and request an audit. He needed to come out and see what we were doing.

Additionally, I worked with the sewer district here. We're on metered water, and we had to pay a sewer discharge fee based on our water use — they didn't know that we were using it as irrigation water, and therefore not going into the sewer. There was a formal application process and a formal audit where they came and saw what we were doing. They said, "We'll still put you on that metered water for the front end, but then on the back end, we aren't going to charge you." So all these ways have reduced our costs.

A big piece of Urban Buds has been availing ourselves of municipal and state grants through the Missouri Department of Agriculture and the city. The city helped us with a facade-improvement program to go from single-paned to double-paned windows. That will help reduce our energy costs. The Missouri Department of Agriculture has been a very good friend of Urban Buds and helped us pay for our high tunnel and our greenhouse.

I am proud that in its first five years, Urban Buds has received a considerable amount in grants, with different cost shares, city, and state programs. For a for-profit farming operation, that's a lot of money. I think urban farmers should consider grant funds as a way to build infrastructure and capacity.

RISK MANAGEMENT

Risk management is an important aspect of keeping your farm business successful over the long term. Managing risk involves identifying, evaluating, and prioritizing risks, then taking actions to reduce those risks or to address your liability. Some of the different risks that farmers may need to address include:

- Weather extremes
- Crop failure
- You or other workers being injured on the job
- Your produce or product making someone ill
- A person being injured while visiting your farm property

These are just examples of the types of risks you assume when starting a farm business. You can be held liable and, if taken to court, could be required to pay damages for some risks, like someone being injured on the farm or someone getting sick from eating your product. This discussion will help you begin thinking about the best ways to minimize your risk and liability so you can farm successfully.

A Process

The process of protecting your farm and assets includes several key steps:

1. **Identify risk exposure.** Where and how can people injure themselves or become ill? Use your imagination and powers of observation to consider all the possibilities.
2. **Analyze exposure.** After you determine possible situations that expose you to risk, evaluate which of them are most serious and require your attention.
3. **Control exposures.** What are the ways you can minimize the most problematic risks? Put up a fence? Post a sign? Eliminate the possibility altogether?
4. **Reduce financial loss.** A financial risk is posed to your business when someone decides to file a lawsuit. Consider how you will pay for the damage. Will you have the cash or capital to assume the full liability? Will you transfer the risk by making insurance policy payments? Will you share the risk by having an insurance deductible?
5. **Implement and monitor.** Establish risk management protocols that protect you and your business. In some agritourism situations, you may require customers to sign a "hold harmless" release prior to paying or visiting your farm or ranch. Retain the proper licenses and permits needed for the business you are operating (refer to Licenses and Regulations earlier in this chapter). Regularly monitor how everything is working and review your policies to make sure you have proper coverage as your business model changes.

First Line of Defense

Perhaps the most important piece of advice when thinking about the safety of your farm is to look at your property with new eyes. As the owner/operator, you may take a lot for granted — that people will know not to climb the ladder that is leaning against the barn or not to pet livestock unless they ask first, or that they know they should wash fruits and vegetables before eating them. *But you cannot assume that others will recognize such risks.* In fact, there are probably risks that you yourself are not aware of since you live and work on the farm every day. If part of your business includes having the public visit your farm, you might want to complete a walk-through with a risk management professional to identify potential hazards on your farm. The goal is to establish an environment that is welcoming and safe.

What are some commonsense actions or strategies you can take to reduce risks on your farm? Do any of the following actions (see box below) apply to you?

IDEAS FOR MINIMIZING LIABILITY RISK

- Store ladders properly after use; don't leave them up for people to climb.
- Keep tractors, implements, and other equipment stored out of sight.
- Provide fruit- and produce-washing stations.
- Fence off areas that are off-limits to visitors.
- Find new homes for animals that show any signs of aggression or unpredictable behavior.
- Provide hand-washing stations for staff.
- Create a personal hygiene and food-safety handbook for staff.
- Properly maintain hand tools and equipment.
- Do safety trainings with staff before they operate equipment.

Educating Your Customers

Any actions you can take to minimize risk and prevent accidents from happening in the first place are going to be helpful. It is also important to educate customers and alert them to possible risks. Here are a few simple ideas:

- At your farmers' market booth, have a sign or remind customers that, while you have washed the produce after harvest, it is always best to wash it again before it is eaten.
- Before a farm tour, have a brief orientation and point out areas that are off-limits or ways to stay safe on the farm. For example, note sections of the farm where there is uneven ground. Or remind visitors that livestock have teeth and hooves, to stay on the proper side of the fence, and to not handle the animals.
- Provide instructions for equipment that people will be using, such as the proper way to manage a bucket when climbing a ladder to pick cherries.

Many farmers find that adults are more difficult to manage than children during farm activities. It may seem counterintuitive, but children are used to following directions at school and other activities and are likely to respect the person giving them information. Adults, in contrast, may take more risks since they feel they have the right to make their own decisions.

Farmers Share Their Experience

RISK MANAGEMENT, BUSINESS STRUCTURES, AND INSURANCE

NOW THAT YOU HAVE LEARNED about some of the risks involved in managing a farm business, how do you feel about the level of risk on your own farm and how to manage it? If you are just starting out, or don't yet have a farm, what level of risk do you think you can be comfortable with? Here, several farmers relate how they have addressed these questions in their own farming operations. Note their perspectives on risk assessment and risk management. If you need information and assistance with this issue, whom might you go to for help?

BLUE FOX FARM

CHRIS JAGGER AND MELANIE KUEGLER, BLUE FOX FARM (profile on page 168)

Chris: There is a lot of crossover between OSHA requirements and GAP requirements. OSHA requirements are just more about keeping your farm tidy and really looking at where people could get hurt, and taking the necessary precautions. Ten years ago, we wouldn't even have thought of managing for those risks. But as you grow and more and more people come on your land, then you really have to look at the issue.

Melanie: Yeah. And you have more of your life invested in the business. One small thing can take it away. So you definitely have to look at how you protect yourself from that.

Chris: We have the family trust that oversees the ownership of the two pieces of land and leases it to Blue Fox Farm. Our main concern has always been about protecting the land. We wanted to make sure the land always remained.

Melanie: We chose to invest in sitting down with a lawyer and figuring out the best way to make sure that the land and the business were taken care of. It was important to us that if something happened, like a lawsuit, that our families wouldn't get involved.

KIYOKAWA FAMILY ORCHARDS

RAINSHADOW ORGANICS

Chris: We've also looked at retirement plans for ourselves, which seems crazy right now, as well as life insurance, so that our children are taken care of in case something happens to us.

CARLA GREEN, SWEET HOME FARMS (profile on page 216)

We have two LLCs. The LLC leases the property from us as owners as a way to manage liability. The LLC is an important model.

We also use farming practices that are less likely to produce food that would cause people to get sick or employees to be injured. We have limited control over the processing plants, and that's always a risk, but that's part of the business.

RANDY KIYOKAWA, KIYOKAWA FAMILY ORCHARDS (profile on page 94)

One of the things that I look at in our direct marketing operation is lowering my risk and liability. That's why I don't allow customers to use ladders in my U-pick orchard and they are instructed not to pick fruit off the ground.

For food safety, I try to make sure that we have clean facilities for both employees at the fruit stand and for customers. We built bathrooms with running hot water and soap.

SARAHLEE LAWRENCE, RAINSHADOW ORGANICS (profile on page 218)

We do a lot of things here that would definitely be associated with extreme liability, so we manage that the best we can.

We have a U-pick garden and we also have volunteers and WWOOFers who come. There's a certain amount of liability in having people's kids here. So those are the two things where we're definitely vulnerable. I guess I just have a lot of faith in people. But other than that, we're trying to facilitate a safe place for people to use some common sense and have a really good time.

BETH HOINACKI, GOODFOOT FARM (profile on page 44)

The farm is set up as a sole proprietorship because it made the most sense for us. I would think setting up an LLC would be pretty important for some folks to protect any personal assets.

Deciding how to structure your farm business is a personal choice; it's important to recognize what are you comfortable with as an individual and what types of assets you need to protect. For us, we weren't going to get a big loan, since we were developing the farm while we had off-farm income. So limited liability didn't really make sense for us. We may have to look at our legal structure at some point.

Liability Insurance

Nobody looks forward to paying insurance premiums. However, if you would like to transfer some of the risk, an insurance policy is an effective way to do it.

Usually when farmers think of protecting their farms, they are thinking about the possibility of buildings burning in fire, loss from theft, or crop failure, and of course there are insurance policies available for these. The USDA's Risk Management Agency offers crop and livestock insurance, for example. General farm and ranch liability umbrella policies covering a certain amount of liability and various activities are also readily available.

As you look for this type of policy, be completely honest and let your insurance agent know exactly what activities or products you plan to offer to customers to be sure they are covered by the policy. Likewise, if you add new activities to your business, such as wagon or pony rides, you should notify your agent to determine if that activity is covered or if your policy needs to be updated. Even when holding a one-time event, it may be important to call your insurance agent beforehand to ask whether or not your policy covers the activities that will take place. In some cases, you can add those activities for a small fee to cover just that event. Your goal is to avoid a scenario where someone is injured and you later find out that your policy does not cover that situation. Such a discrepancy could be devastating to the financial well-being of your farm business.

It can be difficult finding an insurance company or agent who understands farm businesses and whom you trust and feel comfortable with. A first step may include locating other farms in your area that do similar activities as you and asking them who they are insured with. Also check with your state consumer affairs division or insurance commissioner, who may be able to give you tips about how to find an agent and other consumer information.

Part of your risk management plan should also identify whom you will seek legal counsel from, if needed. Some insurance policies cover legal defense if a situation escalates to that, but having an established attorney with whom you work makes good business sense, as he or she would be available to review liability waivers, insurance policies, and other key documents.

Farmers Share Their Experience

OFF-FARM EMPLOYMENT

HOW DO YOU DEFINE SUCCESS? To answer this question, you must take another close look at the vision you have for your farm, your values, and your goals (covered in chapter 1, "Dream It"). Profitability is a key aspect of success, and that means paying yourself (also covered in chapter 4, "Manage It"). But some farmers may hold themselves to an unrealistically high standard, believing, for example, that the farm business should be able to entirely support the family and that off-farm income should not be needed. That assumption needs to be questioned. In the comments here, several farmers describe how off-farm employment is an essential component of their personal and business goals.

GOODFOOT FARM

BETH HOINACKI, GOODFOOT FARM (profile on page 44)

As we have built and supported the farm, we've always had the idea that we don't want to be stressed out about income. We don't like being in debt, so we didn't want to take out loans.

The plan was for me was to stay at home and develop the farm and be the stay-at-home parent for our two kids, while Adam supported the household with a full-time off-farm income. I also work part-time off the farm. Adam continues to work full-time, and he basically pays for our mortgage and the bills, and then provides some capital investment for building the farm.

Our idea is to do what we want. We both want to farm and we both have off-farm interests. The idea is for the farm to provide income, but for us to have enough freedom to pursue off-farm work opportunities that also provide income and are personally rewarding.

So the idea is to balance the farm work, which can be more than full-time work, with those off-farm interests and income. We're not willing to go without health care, which we get from Adam's job. We're not willing to risk that for ourselves or for our kids.

MIRANDA DUSCHACK, URBAN BUDS: CITY GROWN FLOWERS

(profile on page 129)

For us, the goal was for the business to be supporting itself by the fifth year. At that point, we would no longer be dumping our paychecks into it. The fourth year, the business did start supporting itself. But before that, we would take $500 from her paycheck and $500 from my paycheck and put it toward the business. Mimo was able to leave her job with the Cooperative Extension in year 5. I'm still with the Cooperative Extension as of year 7. We really need to build this business so we have some security for our family.

LISA MISKELLY AND ANTON SHANNON, GOOD WORK FARM

(profile on page 92)

Lisa: Pretty much the whole time we've been farming, I have worked at a restaurant between one and four evenings a week and Anton has done odd jobs in the winters. We've definitely needed that additional income to support ourselves. And then I got a full-time job running the student farm at a local college. Having full-time work off the farm is new for us. We felt like we needed to make that shift when we moved to the new farm and got a mortgage. We also took out a loan from the FSA microloan program, so this is really the first time that the farm has taken on debt.

There were personal reasons as well. It's not a sacrifice. I love my job. I'm excited to be working with students, but the vision for our life and for the farm has definitely shifted.

Anton: Since we've taken on debt, having a full-time off-farm job made a lot of sense from a quality-of-life perspective and in taking pressure off of ourselves and the land. We're lucky that Lisa was able to find a job that she likes so much, too, so it's not a sacrifice.

RUNNING A SMALL FARM BUSINESS

Is farming a business or a way of life? If it's a mix of the two, how do you balance those elements? As a business owner and manager, you must be multitalented, with a wide set of skills. As Wendell Berry wrote, "The good farmer . . . must be master of many possible solutions, one of which he must choose under pressure and apply with skill in the right place at the right time." As a small-scale farmer, you must be a problem solver and master of many trades.

In this book, you have explored many of the decisions and details that are involved in pursuing your small farm dreams. But those dreams must be based in reality. To plan and run a small farm business, you must take a close look at some of the personal situations that make it challenging to run a small farm. These include health insurance, retirement, cash flow, and off-farm income. You must also look at the business-related challenges, which often arise as you try to manage your growing business. These include farm size, profitability, and burnout.

Health Insurance

Health insurance for yourself and your family is extremely important. In addition to regular health issues that arise as part of life and aging, farmers are at higher risk for accident and injury than the general population. Some key points to consider:

- If you are an employer, think about paying for group insurance, where part of the premium is paid by the business. Group premiums will be somewhat lower than individual rates, and the same medical screening doesn't always exist in group policies.
- Many associations and organizations, such as the Farm Bureau or a cooperative of farmers, offer health insurance for their members at group rates. If you are a college graduate, check out your alumni association to see if it offers health insurance.
- Health insurance cooperatives, like other cooperatives, are based on the model that the owners of the organization are the consumers of the desired product — in this case, health insurance. The Farmer's Health Cooperative of Wisconsin is an example (see Resources).
- The Affordable Care Act website offers current information on health insurance (see Resources).

Retirement

At different points in this book, you have had the opportunity to think about your vision, mission, and goals.

- Within that vision, how much have you considered what will happen as you get older on the farm?
- What kind of retirement do you envision?
- How much do you want to stay involved in the business?
- Will you want to reduce your hours? (Compared to the rest of the U.S. workforce, farmers tend to work longer than the average person, continuing to work full-time on the farm into their 70s.)
- How much do you need for retirement?
- How do you save money for retirement while also building and reinvesting in the farm business?

Talk to a financial planner now about your retirement goals and develop a plan to reach those goals.

Cash Flow

Cash flow is often the most difficult aspect of farming. Cash flow fluctuates widely — from positive to negative — depending on the time of year. It is important to develop a strategy at the beginning for managing cash flow. You will need to revisit that strategy each year and make changes as necessary. We covered cash flow in more detail in chapter 4, "Manage It." You can revisit that chapter and review the three strategies for managing cash flow outlined there:

- **Advanced payment arrangements:** Establish sales arrangements where you can charge customers in advance, such as in community-supported agriculture.
- **Receivables:** Take steps to get paid quickly; this can be accomplished by setting strict and short-duration payment terms on restaurant and wholesale accounts.
- **Borrowing:** Tight cash flow may require that you utilize loans or short-term credit; shop around for the best terms possible on any borrowing plan.

Off-Farm Income

Most families rely on two incomes to make ends meet. According to the U.S. Bureau of Labor, the number of dual-income families in the United States increased by 31 percent between 1996 and 2006, from 25.5 to 33.4 million families. That number has continued to rise. In spite of this reality, farms are often not recognized as a "real farm" if off-farm income is part of the picture. Somehow the measure of success is that you must make it solely based on income from the farm. It is not clear how this perspective has taken hold, but this kind of pressure is not a reality in other lines of work. If someone owns a restaurant business or small bookstore, are they considered less successful because their spouse works another job as well? Definitely not.

You can still be a sustainable, profitable business if one partner works off the farm. As noted elsewhere in this book, cash flow problems are common when running a farm business, so maintaining other sources of income to support your family and desired lifestyle can be important. However, it is critical that you keep your home and business finances separate and that you make a plan for how you will keep track of on-farm and off-farm income, such as with separate accounts. Off-farm income should not be used to subsidize a failing business; the business needs to pay its own way. On the other hand, if the farm is operating at a net profit and the off-farm income is already covering personal bills, such as health insurance and mortgage, then that is icing on the cake.

Farm Size

When you're trying to determine the right scale for your farm business, you might find yourself in the challenging situation of constantly trying to adjust your size based on the needs and pressures you experience in the business. Some forces may have you thinking you need to expand the business. For instance, economies of scale and efficiency, the need for larger volume of product to sell in a particular market outlet, the need for more income, or the desire to add a new enterprise may have you thinking about growth. Other concerns such as your quality of life, complications of hiring and managing labor, or cash flow problems may be leading you to scale down your operation.

Consider the following scenario: Your business has expanded and you have grown too big to do all the work yourself, so you hire people to help, but then more of your cash flow is going toward paying employees instead of going into your bank account. To correct that situation, you scale down to improve your cash flow, but then you may decrease your quality of life again because you are working too much. What has been your experience with this dilemma?

Profitability

The reality is that as you get started, you will have good years, and you will have bad years when it feels like you are struggling to keep the business going. How do you handle those fluctuations? Many farmers say their vision, mission, and goals help them maintain their perspective through good times and bad. It is also important to recognize how family needs change over time as children grow up and as we age.

As you move along in your farming career, you will weather challenges with more experience. You will be able to home in on the system that works for you and leads to the sort of profitability (and consistency) that you aim for.

The following concepts have been highlighted throughout this book, but they are worth repeating. To be successful and profitable, farmers must:

- Be dynamic.
- Constantly adjust to different market, societal, and environmental changes.
- Be conscious of and revisit family and lifestyle goals to avoid burnout.
- Keep good records so they can monitor their situation and progress.
- Be flexible and open to new ideas.

It will get easier with experience!

Burnout and Balance

In his book *Wisdom of the Last Farmer*, David Mas Masumoto writes, "I often whisper to myself, 'This farm is going to kill me.'" Farming is hard work, and burnout is common. A farm will let you work as hard and as long as you like. It will always have something for you to do. It doesn't care if you work all night. It doesn't care if you work 100 hours a week. It doesn't care if you never take vacations. It doesn't care if you get divorced. It doesn't care if you have any friends. It is up to the farmer to set boundaries.

How do you avoid burnout? First, plan carefully and have realistic expectations. Josh Volk of Slow Hand Farm in Portland, Oregon, planned his urban-scale farm to fit his life and his values. He specifically designed the size and scale of the farm business to avoid burnout and allow him to pursue other activities like writing and consulting. His farm business design allows him to do his farm chores, including CSA deliveries, two days a week. Does this model stimulate some ideas for how you might design your own farm business?

Second, maintain your passion for the business. As you learned in chapter 1, "Dream It," passion won't solve all the problems, but it will give you the energy to work through hard times.

Third, as described in chapter 5, "Grow It," it is important to create a farm environment where the natural cycles, and your crops and livestock, do as much of the work as possible. Working against nature will ultimately result in more work for you as the manager.

Fourth, cultivate strong and supportive relationships with your family and others involved in the farm business. Whether you are a partnership or a family, this will make the work easier and seem less stressful.

Finally, learn to take things slowly and don't get in over your head. Take the time to develop a complete set of farming and business management skills. In the long run, this will pay off both in terms of business success and reduced stress.

10 LESSONS LEARNED

Melissa Matthewson at Barking Moon Farm in southern Oregon provides these 10 lessons for starting and managing a small farm business.

1. Know what you want to make from your farm. Set an income goal and move toward that goal.
2. Take it slow. Don't jump in too fast too soon. This can cause great stress and result in overall failure of the farm business.
3. Plan, plan, plan, and plan some more.
4. Stay in tune with market forces and competition. Be ready to shift every year. Increase your strategies and brace yourself for new competition in the markets. It will happen.
5. Innovate! This is key. Successful farmers are able to innovate on the farm, creating efficiency and success.
6. Build community. Don't farm alone. Other farmers will be a great source of comfort and reassurance. You are all going through the same thing.
7. Don't forget quality of life and values. Know what you want your life to look like and don't forget about that as you begin to farm.
8. Brace yourself for the myriad challenges that lay ahead. Know you will be stronger and able to deal with bumps over time with experience and seasons under your belt. Know that farming is really hard work. But it does get easier with time.
9. Remember you are doing something unique. It is important, valuable, and meaningful work. Seriously, we often forget to tell ourselves this, and when we do, when we remember that we are digging in and producing food for the community and they appreciate it, we can move forward with the hard work of farming.
10. Never undersell yourself, your products, or the work that you do on the farm. Be aware that farming costs a lot of money. Buying implements, infrastructure, seed — all of these things are a huge output of capital, especially in the first five years, so accordingly you need to charge the right price for what you produce on the farm. Don't be shy to ask for a higher price than other farmers. If the quality is there, you will have customers.

Farmers Share Their Experience

CHALLENGES AND ADVICE

HOW DO YOU DEFINE SUSTAINABILITY? We focus a lot on the sustainability of the farm. What about the sustainability of our own lives, our health, and our families? Consider these heartfelt observations from farmers who have been there, in the trenches, working day to day, year in and year out, trying to create a successful business. As you read, take a few notes and write down what has helped them get through the challenging times and what keeps them going.

CARLA GREEN, SWEET HOME FARMS (profile on page 216)

Staying sane when you have a seven-day-a-week job that's pretty much every waking minute is a struggle.

We were trying to run our business out of the kitchen, and it was a nightmare. We had literally stacks of everything on the counters and on the desks, and it was horrible. So we rented a mobile office building, like a construction office building. It has one room for an office and one room for a conference room. Having a place to organize things and go for meetings has helped enormously. We struggle to separate work life from non–work life so that we're not working 100 percent of every day all the time, except when we're sleeping.

URBAN BUDS

JOSH VOLK, SLOW HAND FARM (profile on page 46)

There tends to be burnout on full-time farms. I had a lot of success developing a three-acre farm for somebody that was run entirely by folks working part-time. I didn't see any burnout in people because they were doing other things off the farm and bringing those experiences to the farm. There was just a lot of energy. I wanted to do this one as a part-time farm so that I would have time to do some other projects within agriculture that I was interested in.

MIMO DAVIS AND MIRANDA DUSCHACK, URBAN BUDS: CITY GROWN FLOWERS (profile on page 129)

Mimo: For the first five years, we were here all the time. Now we're trying to take Sundays off and we're pretty successful at that.

Miranda: We are not a nonprofit. We have a son. We're trying to do this. And I think we can do it in a way that's still authentic. On my communal farming projects, where profit was not the goal, it was really easy to sleep in if you were tired. It was really easy to go to that concert or head out to Burning Man or whatever.

Mimo: No, we're about profit. We have to wake up.

CHRIS JAGGER AND MELANIE KUEGLER, BLUE FOX FARM

(profile on page 168)

Chris: The whole farming movement is all about soil health and farm health. I think a lot of times people forget that they, the people, also need to be healthy. The fact that we need to focus on emotional sustainability as well as soil sustainability was a big eye-opener for us.

Melanie: And make sure that we get a little bit of time off. Sundays are really important to us. That is our day to be with the family. Nothing stops that.

Chris: We often ask elder farmers for advice. There's an amazing network of older, more experienced farmers. They have told us that you will never learn everything. And if you think that you have, then you're wrong.

RANDY KIYOKAWA, KIYOKAWA FAMILY ORCHARDS (profile on page 94)

The advice I have is if you love it, know why you're doing it. It's not easy. You're going to have to wake up early, go to sleep late. You've got to be out there in the hot weather and the cold weather.

BETH HOINACKI, GOODFOOT FARM

(profile on page 44)

This summer was not what I thought it was going to be. We didn't get our hoop houses up. We couldn't get into the ground when we wanted to because we had an irrigation project that stalled out and finished behind schedule.

At one point this summer, I had this "aha" moment — I realized I was stuck on this idea of how it's supposed to be instead of going with the reality of how things were. Later that day, I said to my husband, Adam, "I've got it finally figured out. It's never exactly the way that I think I want it to be. But it's always just good enough. And that's good enough for me."

RENARD TURNER, VANGUARD RANCH (profile on page 132)

To farmers who are thinking about doing this, I would say that economy of scale is the most important thing. The smaller you are, the smarter you have to be. It certainly is not a get-rich-quick scheme. It doesn't work that way at all. It's based on proximity to demographic centers. If you're closer to a city with a large population, you're going to have more options for selling your product. You have to distinguish yourself by doing something superb and/or unusual, or superbly unusual, in order to make a niche in this market. There's lots of competition.

There are a lot of people who have no agricultural background. I don't want to discourage them, but you really have to become a student of the craft. You really have to read and study a lot, but there's no substitute for experience.

I didn't start out to do exactly what I'm doing now. It was a painful process of discovery. But as one thing didn't work out, the next thing did. We tried a number of things. You have to have a lot of faith, and well, if you don't have faith, you're not going to last. So either your faith is going to strengthen, or your pocketbook's going to weaken. That's pretty much how it works.

You need a lot of support to farm. You need a good partner, and it helps to have a fair amount of money to start. I did not have a lot of money to start, so I did a lot of work myself. If you're willing to do the work, it can be very rewarding, but there are lots of challenges to making it profitable. Some years are good, and some years are bad. That's the reality of it.

You can do this, but you're going to have long hours, you're not always going to make money, and you'll be up early in the morning. But for those of us who can farm, it's a great life. It gives you a chance to eat better than most people. And it gives you a sense of pride that is hard to beat.

VANGUARD RANCH

The 12 authors are part of the Center for Small Farms & Community Food Systems at Oregon State University or in other educational capacities in farm production, food system development, and family and farm business management. Our approach to all our research and educational programs and this book is grounded in our partnerships with farmers. Sometimes we call this "the farmer and the egghead." The strength of this approach is seeing the big picture and understanding the "system" and its linkages on one hand, and articulating the ways specific farms put all of that into practice on the other.

GARRY STEPHENSON is Oregon State University Extension Small Farms specialist and Director of the Center for Small Farms & Community Food Systems. He has over 30 years of experience working with small farmers on educational programs and applied research. Over the years, his work has focused on alternative marketing, sustainable farming practices, beginning farmer education, and the production and policy needs of organic farmers. Garry coauthored the introduction, "Growing a Farm," plus chapter 1, "Dream It," chapter 2, "Do It," chapter 3, "Sell It," chapter 5, "Grow It," and chapter 6, "Keep It."

NICK ANDREWS is an Associate Professor of Practice with the Oregon State University Extension Small Farms and Organic Extension Programs. His work focuses on organic vegetable production, cover crops, nutrient management, and pest management. He was previously an organic farm and processing inspector with Oregon Tilth, an independent IPM consultant in Hood River, and vegetable pest management consultant with the UK's Agricultural Development and Advisory Service. Nick coauthored chapter 5, "Grow It."

DAVID CHANEY is a lead consultant for DEC Education Services. He works nationally with universities and nonprofit groups to design and produce innovative educational resources to enhance learning that goes beyond information, to knowledge and action. He has more than 30 years' experience developing educational programs and materials related to sustainable agriculture. David coauthored chapter 5, "Grow It."

MELISSA FERY is an Associate Professor of Practice with the Oregon State University Extension Small Farms Program. She has been working to support Oregon agriculture in the areas of soil quality and land management for over 20 years. Melissa and her family live on a small acreage where she enjoys a rural lifestyle growing plants and raising animals. Melissa coauthored chapter 1, "Dream It," and chapter 6, "Keep It."

AMY GARRETT is an Assistant Professor of Practice with the Oregon State University Small Farms Program. She has more than 20 years of experience in the horticulture industry ranging from landscape design, installation, and maintenance to organic farming, research, and education. Drought mitigation tools and strategies for growing with little or no irrigation are a current focus of her work. Amy coauthored chapter 3, "Sell It," and chapter 5, "Grow It."

LAUREN GWIN is an Oregon State University Extension Community Food Systems Specialist and Associate Director of the Center for Small Farms & Community Food Systems. For the last 15 years, her work has largely focused on infrastructure, value chains, and public policy related to sustainable agriculture and local and regional food systems. In the last 5 years, her work has increasingly focused on the intersection of farming, food, health, and equity. Lauren coauthored the introduction, "Growing a Farm," and chapter 6, "Keep It."

MELISSA MATTHEWSON owns and operates Barking Moon Farm with her husband near Applegate, Oregon. She is former Oregon State University Extension Small Farms faculty. She is the author of a forthcoming book of nonfiction. Melissa coauthored chapter 2, "Do It," and chapter 6, "Keep It."

TANYA MURRAY is an Organic Education Specialist with Oregon Tilth, Inc. She has over 10 years of experience working on and managing organic vegetable farms. Her work is focused on developing and delivering programming that supports farm viability. Tanya has an MBA from Portland State University and a bachelor's degree in Education for Sustainable Agriculture from Prescott College. Tanya coauthored chapter 4, "Manage It."

HEIDI NOORDIJK is an Outreach Program Coordinator with the Oregon State University Extension Small Farms Program. She has over 15 years of experience in applied research and education for farmers, and 10 years of experience working on and managing farm crews on small farms and not-so-small orchards. Her current focus is integrated pest management, sustainable farming practices, beginning farmer education, and farmer networks. Heidi coauthored chapter 5, "Grow It."

SHERRI NOXEL is the Director of the Austin Family Business Program at Oregon State University. Sherri designs educational programs and networking opportunities for Oregon's multigenerational businesses and family business students. Sherri has an Advance Certificate in Family Business Advising from the Family Firm Institute and was the leader of two USDA projects to support families in transitioning farm management responsibilities. Sherri coauthored chapter 6, "Keep It."

MAUD POWELL is an Associate Professor of Practice with the Oregon State University Extension Small Farms Program. She and her husband own and operate Wolf Gulch Farm in the Siskiyou mountains of Southern Oregon, where they grow certified organic seeds and winter vegetables and manage the Siskiyou Sustainable Cooperative's seven-farm CSA program. Maud coauthored chapter 2, "Do It," and chapter 5, "Grow It."

JOSH VOLK is the proprietor of Slow Hand Farm in Portland, Oregon, and has been working on and managing small farms around the United States for the last 20 years, studying the systems that make them efficient. He travels around the world, consulting with farmers and researchers, teaching new farmers, and presenting workshops. Josh coauthored chapter 4, "Manage It."

Contributors:

TYE FOUNTAIN, Pacific Ag Systems, Inc., contributed to chapter 2, "Do It."

TODD MOSS, Syracuse University, Syracuse, New York, contributed to chapter 6, "Keep It."

KRISTIN POOL, Kristin Pool Productions, Portland, Oregon, contributed to the introduction, "Growing a Farm."

APPENDIX
THE FARM PLAN

To assist you in developing your own whole farm plan, we include a whole farm plan template — 23 separate worksheets (appendices 1-1 through 1-23) referenced throughout the book and identified by the Farm Plan icon. Where this icon appears in the text, you will find a brief description of the worksheet, plus instructions or suggestions for completing the form. Further details and instructions appear on the following worksheets. Assemble these worksheets and other supporting documents into a binder or folder. **You can also find these worksheets online, at www.storey.com/farm-plan-worksheets**.

APPENDIX 1-1

FARM TEAM SKILLS ASSESSMENT

What skills do you and your farm team have in each category? Write the name of the farm team member and their skills.

What skills do you need to develop or get help with? List skills that you do not have within your farm team, and state how you plan to meet that need (e.g., training, hire help).

TASKS/SKILL AREAS	SKILLS YOUR TEAM HAS	SKILLS YOU NEED
Planning (e.g., research, resource assessment, lead meetings, writing)		
Business management (e.g., organizational, communication, financial, bookkeeping, record keeping, payroll, tax prep, inventory)		
Labor management (e.g., communication, training, supervision, scheduling)		
Equipment and buildings (e.g., construction, mechanical work, engine work, repair, maintenance)		
Crop management (e.g., soil management, drive tractor, irrigation, crop production practices, cropping systems, harvest and storage)		
Livestock management (e.g., animal health and reproduction, grazing management, feed and nutrition, slaughter and processing)		
Marketing (e.g., communication, market research, planning, contracting)		

APPENDIX 1-2

QUALITY-OF-LIFE ASSESSMENT

Owning and operating a farm offers a unique quality of life. As part of your whole farm planning process, think about the answers to the following questions. Suggestion: Have your farm partners answer these questions separately, then discuss your answers.

Do you like to mix your personal life and work life or keep them separate? Why?

____

____

____

How valuable is having leisure time with friends and family?

____

____

____

How much do you like working with others, including employees, family members, and business partners? Would you rather work by yourself?

____

____

____

Do you enjoy marketing and having contact with customers? Why or why not?

____

____

____

What are your favorite tasks on the farm (e.g., handling animals, production, or marketing)?

____

____

____

Approximately how many hours a week are you willing and able to work? Consider both on-farm and off-farm work.

Are you a risk-taker? Are you comfortable with uncertainty, or do you prefer to know what to expect in most situations? Why?

What does the phrase "financial security" mean to you?

What are your family members' goals and interests? How do they align with yours?

What other demands are made on your time? Consider family, health, hobbies, and other time commitments.

Would you prefer to have family members perform all farm labor, or are you interested in hiring outside help?

APPENDIX 1-3
IDENTIFYING YOUR FARM: NAME AND CONTACT INFORMATION

Farm name. Your farm name may describe what you do, indicate where you are located, reflect your personality, or enhance marketing of your products. Write down your farm name, or use the space provided to brainstorm possible names.

____

____

____

Farm team. List everyone who is involved in the farm business — family members, employees, seasonal labor, interns, and so on.

Owners/operators: ______________________________

____

Other members of your farm team: ________________________

____

Contact Information

Location address: ______________________________

Mailing address: ______________________________

____

County: __________________________________

Phone number(s): ______________________________

Fax number: ________________________________

E-mail address: ______________________________

Website: __________________________________

APPENDIX 1-4

IDEAL FARM

When you think about your current or future farm, what do you picture? Write a brief description of your ideal farm.

As you finalize your whole farm plan and put all the pages together in a binder or folder, you may want to add to this description with a hand-drawn sketch of the farm or a few photos. You can insert those here as separate pages.

APPENDIX 1-5
VALUES AND VISION

Values

Values are core beliefs and philosophies that reflect your views on life. They often influence your goals and business decisions and help guide management of your farm. Values typically do not change with time and are reflected in everything you do. List some of your values.

Vision

A vision statement describes the big picture of your business over time. It defines an ideal future and the impacts on your local community or society in general. Your vision may include what you want your farm to look like in 10 years, what products you'd like produce, or how your farm will grow. Write a draft vision statement. This is something you can revisit and update as your circumstances and situation change.

APPENDIX 1-6

ENTERPRISE SELECTION: ANNUAL CROPS

Annual cropping systems include vegetables, grains, legumes, oilseed crops, herbs, and flowers. These are often grown as primary cash crops on a small farm or as part of an integrated farm. Challenges of annual cropping systems include pest, weather, and disease problems as well as labor and marketing. Annual crop production becomes especially labor intensive during spring and summer. Most annual crops are marketed directly through farmers' markets, farm stands, retail stores, restaurants, community-supported agriculture (CSA), and U-pick operations. The benefits of annual cropping systems include steady annual cash flow and high-value return.

Answer the following questions to assess your readiness for annual crop farming and what type of crop mix might best suit your situation.

How will you accommodate the rigorous schedule that is required throughout the growing season for annual crop systems?

___

___

___

Many annual crops are sold through direct marketing channels. Will you enjoy interacting with customers?

___

___

___

If you decide to market through wholesalers, which usually brings a lower price, can you grow enough to sustain a business at wholesale prices?

___

___

___

Some farmers concentrate on one or two annual crops; others operate diversified cropping systems. What type and how many varieties of crops do you want to grow?

Diversified vegetable production is complex, fast paced, and intense for much of the growing season. Does this fit with your personality, physical capabilities, and lifestyle?

Annual crops can be grown on rented ground. Is this an option that would help you get started or expand production?

Will your soil and water rights support annual crop production?

Do you have the crop management skills and experience needed to grow the crop? If not, how will you ensure the success of your farm, especially when you are just starting out?

APPENDIX 1-7

ENTERPRISE SELECTION: LIVESTOCK

Livestock systems include pork, lamb, goat, beef, dairy, broiler, and egg production. Value-added products include milk, cheese, and many processed meat products. Livestock can be raised in a multispecies system to enhance production and marketing opportunities. Livestock also may be part of an integrated crop and livestock farm and a good option for land that isn't suited to growing crops. Livestock need care and attention year-round. Animals require shelter, adequate nutrition and water, safe fencing, exercise, and health care. Proper pasture management is an essential aspect of sustainable livestock production for many small farms. State and local regulations and other legal parameters can affect what and how livestock can be raised on a particular site. Investigate these issues before initiating a livestock enterprise.

Answer the following questions to assess your readiness for raising livestock and what type of livestock business might best suit your situation.

Do you enjoy handling animals and doing daily chores?

____

____

Can you care for sick and injured animals? Are you capable of giving injections?

____

____

Are you willing to feed livestock on a regular schedule when pasture is not available? If you own dairy animals, are you willing to milk twice a day? If you get sick or are unavailable, can someone else fill in?

____

____

____

Are you willing and able to go out in freezing weather and break ice on the water trough or help deliver a newborn calf?

____

____

____

Are you aware of and able to perform routine management practices such as castration, dehorning, tail docking, feet trimming, and wing clipping?

What is the purpose of your livestock enterprise (e.g., meat, fiber, milk, or multipurpose)?

Are you willing to send animals to slaughter?

Do you have a replacement livestock plan?

Do you own, rent, or have access to enough range or pasture for the number of animals you need to raise to be profitable? Is water available?

Do you have adequate shelter, fencing, and handling facilities?

Are you interested in commercial livestock production, breeding stock production, or both?

Where will you obtain foundation breeding stock?

Will you use natural mating or artificial insemination for breeding? What type of breeding program will you implement?

Do you have access to a knowledgeable veterinarian who will provide service to your farm?

How will you manage livestock manure? Manure is a resource, but it can also be a problem if not managed properly.

How will you dispose of dead animals?

APPENDIX 1-8

ENTERPRISE SELECTION: PERENNIAL CROPS

Perennial cropping systems include nut and fruit trees, cane berries, hops, grapes, and grass seed. These are often grown as primary cash crops or as part of an integrated farm. Perennial and annual cropping systems have different nutrient and pest management needs and require different cultural practices.

Answer the following questions to assess your readiness for managing perennial systems and what type of crops might best suit your situation.

Do you have enough land to make the business profitable? For the crop you are interested in, how big of an area is needed and how many plants are necessary?

Do you have the finances for the initial investment required in establishing the orchard, vineyard, et cetera?

How long will it take for the crop to produce yields suitable for commercial production? Can you survive financially while you are waiting for the crop to begin producing? What is your plan for doing so?

How long will the crop be productive?

Some perennial crops require annual pruning, which is often completed during winter. Are you willing to spend considerable time outside in cold weather, possibly climbing up and down a ladder?

Do you have the crop management skills and experience needed to grow the crop? If not, how will you ensure the success of your farm, especially when you are just starting out?

APPENDIX 1-9

MISSION AND GOALS

Mission Statement

A mission statement is a set of guiding principles based on your vision and values. It describes the overall purpose of your business and may include what you do, how and why do you do it, and who you want to serve. Mission statements can sometimes be used as marketing tools. Periodically review your mission statement and update it if necessary to keep your business dynamic. Write down your mission statement or some key points that you will include in your mission statement.

____

____

____

____

____

____

____

____

Goals

Goals can be categorized as short, medium, and long term. Goals should align with your values, vision, and mission. List your major goals in the space provided. As you write your goals, keep in mind the SMART acronym discussed in chapter 1, "Dream It": goals should be specific, measurable, attainable, relevant, and timely.

Medium-term and long-term goals: ______________________________

____

____

Short-term goals: ______________________________

____

____

APPENDIX 1-10
LAND RESOURCE INVENTORY

Legal description of farm: ______________________________

Township: ______________ Range: ______________ Section: ______________

LAND	ACREAGE	OWN OR RENT	WATER RIGHTS	COMMENTS
Tillable land				
Pasture				
Woodlot				
Other				

APPENDIX 1-11

SOILS ASSESSMENT

Use the information you obtained from the online Web Soil Survey (see Resources) to complete the table below. If you cannot access the Web Soil Survey, or you prefer working with print copy, visit your local soil and water conservation district or Cooperative Extension office to look at a copy of your county's soil survey. If you do not yet have farmland, consider what types of soils and soil properties you will need for your ideal farm and fill the page out accordingly.

SOIL SERIES	CAPABILITY CLASS	DRAINAGE	EROSION RISK	OTHER CHARACTERISTICS	LIMITATIONS AND OTHER ISSUES

APPENDIX 1-12
WATER RESOURCES ASSESSMENT

If you have questions about your water rights or whether your property has water rights, contact your state water resources department. To identify health and food concerns, it is important to have irrigation and drinking water tested for bacteria, nitrate, arsenic, salts, pesticide residues, and other water quality issues. List agricultural and domestic water sources available on your farm; the use and delivery system associated with each; and notes about quantity, quality, and limitations.

SOURCE	USE	DELIVERY SYSTEM	QUANTITY, QUALITY, LIMITATIONS
Example: Surface water rights from Thomas Creek	*Irrigation for 6.5 acres*	*Overhead sprinklers on movable hand line*	*Junior water rights; water may be limited in drought years*

Are there state or county restrictions on the use of domestic well water for agricultural purposes?

Are there state, county, or local restrictions on the use of gray water or rainwater collection?

APPENDIX 1-13
SWOT ANALYSIS

Remember: strengths and weaknesses are internal to the business; opportunities and threats are external.

Strengths	Weaknesses
Opportunities	Threats

APPENDIX 1–14

INFRASTRUCTURE AND EQUIPMENT ASSESSMENT

Enterprise: ______________________________

(e.g., peaches, peppers, cheesemaking, etc.)

INFRASTRUCTURE

RESOURCES CURRENTLY ON THE FARM OR RANCH	WHAT WE NEED	HOW TO GET IT (acquisition strategies)

Enterprise: ______________________________

(e.g., peaches, peppers, cheesemaking, etc.)

EQUIPMENT

RESOURCES CURRENTLY ON THE FARM OR RANCH	WHAT WE NEED	HOW TO GET IT (acquisition strategies)

APPENDIX 1-15
LABOR PLAN

Year 1

List your major goals regarding labor management on your farm or ranch. Consider the issues covered in chapter 2, "Do It," such as type of employees (seasonal or permanent), how many employees, importance of having year-round employees, relationship with employees, education and training of staff, becoming a good manager, etc. *For each goal, write a brief statement of how you plan to achieve that goal.*

GOALS	STRATEGIES FOR MEETING GOALS
1.	
2.	
3.	
4.	
5.	
6.	

INITIAL/CURRENT LABOR SETUP

In the table below, describe the current labor configuration for your farm or ranch. If you are just starting out, describe as best you can what kind of labor setup you think you will need to meet your goals. Think about how you will find employees and/or seasonal labor, how they will become part of your farm team, and how you plan to meet labor costs and expenses.

EMPLOYEES	COMPENSATION PACKAGE (Salary, Pay Rate, Taxes, Benefits)
Farm team (regular staff/employees)	
Seasonal labor	

5-Year Projection

CROP OR LIVESTOCK PRODUCTION ACTIVITIES

Do you plan to expand your existing enterprises (e.g., add more acres, increase number of animals)? Do you plan to add any new enterprises?

If yes to either of these questions, describe those changes in the space below. Also note any marketing innovations or value-added activities (e.g., food processing) that would affect your labor needs.

LABOR SETUP REQUIRED TO REACH GOALS FOR NEW OR EXPANDED ENTERPRISES

In the space below, describe what labor configuration you will need to account for any new or expanded business operations. As you did with the initial/current configuration, consider how you will find employees and/or seasonal labor, how they will become part of your farm team, and how you plan to meet increased labor costs and expenses. In the compensation package column, note any salary or per-hour increases due to pay raises.

EMPLOYEES	COMPENSATION PACKAGE (Salary, Pay Rate, Taxes, Benefits)
Farm team (regular staff/employees)	
Seasonal labor	

PROGRESS TOWARD MEETING YOUR LABOR GOALS

Make a point each year of revisiting the goals you set for yourself at the beginning of this worksheet. How are doing in reaching those goals? Do you need to modify any of them? Do you need to make and adjustments in your strategies?

APPENDIX 1-16
SIMPLIFIED FARM ENERGY ASSESSMENT

BASELINE FARM ENERGY USE

SUPPLY SOURCE	CURRENT USE (BTU)	PLANNED USE (BTU)	NET SAVINGS ($)
Electricity [1]			
Natural gas [2]			
Propane [3]			
Diesel [4]			
TOTAL			

[1] 1 KWh = 3,410 Btu
[2] 1 therm = 100,000 Btu
[3] 1 gallon propane = 91,000 Btu
[4] 1 gallon diesel = 139,200 Btu

PLANNED ENERGY USE REDUCTION STRATEGIES

FARM ACTIVITY	CURRENT ENERGY USE*	CONSERVATION/ENERGY REDUCTION STRATEGY
Tractor (tillage, mowing, and so on)		
Truck/other vehicles		
Irrigation pump		
Heating and cooling buildings		
Heating and cooling product		
Refrigeration/freezing		
Drying		
Dehumidifiers for storage		
Lights		
Fans		
Milking machines		

*Indicate actual amount if known, or relative usage — low, medium, high.

APPENDIX 1-17

INFLUENCE OF PERSONALITY AND LIFESTYLE GOALS ON MARKETING

As discussed in chapter 3, "Sell It," small farms need to take advantage of niche markets and high-margin sales in order to optimize income. Direct sales to customers are one way to achieve this goal. Direct marketing options include farmers' markets, community-supported agriculture, farm stands, U-pick, and agritourism. Before engaging in direct marketing, evaluate what you enjoy, your comfort zone with direct sales, and the time you are willing to commit to this form of marketing.

Answer the following questions for yourself, and for other farm team members.

Are you a sociable person? Do you enjoy visiting with people and sharing your story, or would you rather grow and deliver your product without interacting with customers?

____

____

____

How well can you describe and sell your product? What is the story of your farm and your product?

____

____

____

Do you enjoy getting up early to harvest crops, then spending the rest of the day in town selling at a farmers' market?

____

____

____

Do you like to work without interruption, or would you enjoy visiting with people who stop by your farm to make purchases?

____

____

____

APPENDIX 1-18
MARKETING STRATEGY COMPARISON

All marketing channels have advantages and disadvantages. Write down any marketing strategies you are already involved in, and any others you are considering, in the table below. Use the following list as a reference. Evaluate each strategy in terms of its advantages and disadvantages to you. Potential marketing channels include:

- Wholesale
- Roadside stands
- Farmers' markets
- CSA
- U-pick
- Agritourism
- Restaurants
- Public institutions (e.g., schools, hospitals)
- Farmers' cooperatives
- Web/online sales

MARKETING CHANNEL	ADVANTAGE	DISADVANTAGE

APPENDIX 1-19
BUDGETS AND VARIANCES

A budget shows you the costs of production and the expected revenue. Through the budgeting process, farm businesses can evaluate variable and fixed costs as they relate to the profitability of planned production systems. Farmers can apply this information to create projections and plans for income and expenses before the season starts and to track variations from these budget projections through the year.

Using the template below, prepare a whole farm budget for your business. Remember to clearly define the parameters and goals of your budget (type of farm, product, acreage, start-up, seasonal, expansion, and so on). If you are in the very beginning stages and are not yet operating a farm business, create a hypothetical budget based on your current plans and projections.

Budget for: ______________________________

EXPENSES	AMOUNT ($)
Variable	
Fixed	
TOTAL EXPENSES	

INCOME	AMOUNT ($)
TOTAL INCOME	
Profit/loss = Total income – Total expenses	

APPENDIX 1-20

CASH FLOW AND ACCOUNTING SYSTEMS

Cash Flow

Cash flow indicates how money is received and spent by the business. Planning for cash flow and recording cash flow are important ways for the farm to make sure it has sufficient funds to cover monthly costs, including loan payments.

What strategies are you implementing (or do you plan to implement) to maintain a positive cash flow and avoid cash flow problems?

____

____

____

____

Accounting Systems

All farm businesses need systems and procedures for maintaining accurate records of income and expenses by relevant categories.

What kind of accounting system do you have, or do you plan to set up, for your farm business?

____

____

____

____

____

____

Who will do the accounts and books for the business? ____________________

____

APPENDIX 1-21
PRODUCTION COSTS

Farmers need to recognize the full costs of production and overhead and make sensible choices around production costs and pricing. Setting a product price that is competitive with other businesses and that covers expenses provides a livelihood to the farmer.

Fill out the following table for your farm business. Take your time as you think through this worksheet and return to it as needed and as you gather more information.

TYPE OF PRODUCTION COST	AM I MANAGING THESE COSTS WELL?	IF NOT, WHAT STRATEGIES CAN I USE TO CONTROL THESE COSTS?
Equipment		
Infrastructure		
Repairs & maintenance		
Inputs & feed		
Marketing		
Labor		
Administrative		

APPENDIX 1-22

FINANCING

Farmers generally invest a large amount of savings or borrowed money in their business. It is important to examine the pros and cons of different types of financing and obtain adequate and appropriate levels of capital to start and sustain the farm business.

The different sources of financing discussed in chapter 4, "Manage It," are listed in the following table.

If you are farming now: In chart below, indicate if you currently use each source of financing, how much debt you currently hold in that category, and when you hope to pay off that debt. Also indicate if you want to pursue that source of financing in the future.

If you are not yet farming: In the chart below, indicate if you want to pursue that source of financing and estimate how much financing you will seek.

SOURCE OF FINANCING	CURRENT DEBT AND EXPECTED PAYOFF DATE	PLAN TO PURSUE THIS SOURCE OF FINANCING IN FUTURE AND HOW MUCH
Self-financing		
Family/friends		
Commercial bank loans		
Government loans		
Economic development agency or SBA loan		
Government grant or subsidy		
Individual development account		
Business funding		
Credit card		

APPENDIX 1-23

BALANCE SHEETS

Balance sheets measure the wealth of a business at a specific point in time and allow you to analyze how your financial position changes on an annual basis.

Create a balance sheet for your farm business using the template below. If you are already farming, use current figures for your farm. If you are not yet farming, try developing a hypothetical balance sheet using the best data you have.

FARM BALANCE SHEET

As of (date): ____________________

ASSETS	AMOUNT ($)
Current	
Cash in checking	
Cash in savings	
Other investments (CDs, brokerage accounts, and so on)	
Accounts receivable	
Inventory	
Other	
Noncurrent	
Life insurance	
Retirement accounts	
Tools and equipment	
Vehicles	
Land/real estate	
Total Assets	
LIABILITIES	**AMOUNT ($)**
Current	
Current bills	
Short-term debt	
Accounts payable	
Taxes	
Noncurrent	
Long-term debt (mortgages and other long-term loans)	
Total Liabilities	
Net Assets = Total Assets - Total Liabilities	

APPENDIX 2-1
TRACTOR NEEDS: ANNUAL CROPPING SYSTEMS EXAMPLE

	TASK/FUNCTION	FEATURES TO LOOK FOR WHEN BUYING A TRACTOR
Essential	Cultivate soil	Adequate horsepower to pull plow and/or appropriate-size tiller
	Haul equipment and produce bins	Trailer, palette forks
Additional Components If/When Budget Allows*	Weed control	Implements appropriate for your row configuration
	Specialized tillage for cover crops	Low gears and power takeoff (PTO) for spaders
	Produce compost	Front-end loader/bucket for turning compost
	Fertilize field	Drop spreader

*If these are necessary functions and you don't have the finances to acquire the needed equipment, consider renting/leasing the equipment, or look for opportunities to share equipment with other producers.

APPENDIX 2-2
TRACTOR NEEDS: GRASS-BASED LIVESTOCK SYSTEM EXAMPLE

	TASK/FUNCTION	FEATURES TO LOOK FOR WHEN BUYING A TRACTOR
Essential	Haul hay and water to livestock	Trailer
	Mow pasture	PTO + mower
	Manure management	Front-end loader, spreader
Additional Components If/When Budget Allows*	Hay production	Bailer
	Fertilizer/lime application	Spin spreader

*If these are necessary functions and you don't have the finances to acquire the needed equipment, consider renting/leasing the equipment, or look for opportunities to share equipment with other producers.

APPENDIX 2-3

TRACTOR NEEDS: PERENNIAL CROPPING SYSTEMS EXAMPLE

	TASK/FUNCTION	FEATURES TO LOOK FOR WHEN BUYING A TRACTOR
Essential	Work between rows	Appropriate wheel base for maneuvering between rows, low center of gravity for better stability on sloped land
	Cultivation	Specialized cultivation implements to minimize damage to trees/vines
	Haul equipment and produce bins	Trailer, palette forks
Additional Components If/When Budget Allows*	Mow	PTO + mower attachment
	Dig planting holes	PTO + augur attachment
	Spray orchard for disease control	Spray tank and sprayer

*If these are necessary functions and you don't have the finances to acquire the needed equipment, consider renting/leasing the equipment, or look for opportunities to share equipment with other producers.

APPENDIX 2-4
TRACTOR NEEDS

Farm Name:

	TASK/FUNCTION	FEATURES TO LOOK FOR WHEN BUYING A TRACTOR
Essential		
Additional Components If/When Budget Allows*		

*If these are necessary functions and you don't have the finances to acquire the needed equipment, consider renting/leasing the equipment, or look for opportunities to share equipment with other producers.

APPENDIX 3-1

FAMILY BUSINESS ROLES ACTIVITY

For each family member or farm team member, put a check mark in the box for the different roles that person plays.

	ROLES							
PERSON	**MANAGER/EMPLOYEE**		**FARM PARTNERS**		**FAMILY**			
	Manager	**Employee**	**Owner**	**Investor**	**Off-farm income-health benefits**	**Manage household**	**Childcare**	**Other**
You								
Your spouse/ partner								
Siblings								
Children								
Other relatives								
Other farm team members								

SOURCE CITATIONS

2: Do It

Page 82, Labor Laws: "Checklist for Hiring Employees" (Rogue Farm Corps) in *Western Sustainable Agriculture Research & Education (SARE) Farm Internship Curriculum and Handbook*. Powell, Tom, Maud Powell, and Michael Moss. 2010.

3: Sell It

Page 104, How to Market/Four Ps: Perreault, William D., Joseph P. Cannon, and E. Jerome McCarthy. *Basic Marketing: A Marketing Strategy Planning Approach.* 19th ed. McGraw-Hill Education, 2014.

Page 105, How to Market/SIVA: Chekitan, S. Dev and Don E. Schultz. "In the Mix: A Customer-Focused Approach Can Bring the Current Marketing Mix into the 21st Century." *Marketing Management* 14 (no. 1, 2005): 16–22.

Page 128, Sell It: Keys to Success: Stephenson, James. *Ultimate Small Business Marketing Guide*. Entrepreneur Press, 2007.

5: Grow It

Page 172, quote: Coleman, Eliot. *The New Organic Grower: A Master's Manual of Tools and Techniques for the Home and Market Gardener*. 2nd ed. Chelsea Green, 1995.

Page 174, Planning and Intervention Practices for Ecologically-Based Pest Management: Madgoff, Fred, and Harold Van Es. *Building Soils for Better Crops*, 3rd ed. Sustainable Agriculture Research and Education, 2009.

Pages 178 and 183, The Nitrogen Cycle adapted from Agricultural Ecosystem Management, Course 3 in the National Continuing Education Program (USDA SARE); and C. Johnson, G. Albrecht, Q. Ketterings, J. Beckman, and K. Stockin. *Nitrogen Basics—The Nitrogen Cycle*, Cornell University Cooperative Extension, 2005.

Pages 181 and 203, Key Practices in Sustainable Agriculture: The ecosystem concepts and biogeochemical cycles in this section are adapted from Agricultural Ecosystem Management, Course 3 in the National Continuing Education Program, Sustainable Agriculture Research and Education (SARE), USDA National Institute of Food and Agriculture (NIFA). By permission.

RESOURCES

1: Dream It

Online Publications and Videos

Determining Soil Texture by Hand
Organic Farming Systems and Nutrient Management
Washington State University
https://puyallup.wsu.edu/soils/video_soiltexture

Estimating Soil Texture
Organic Farming Systems and Nutrient Management
Washington State University
https://s3.wp.wsu.edu/uploads/sites/411/2014/12/Paper_SoilTextureDiagram.pdf

Northeast Small Scale "Sustainable" Farmer Skill Self-Assessment Tool
New England Small Farm Institute
www.smallfarm.org/uploads/uploads/Files/indebthselfassessment.pdf

Websites

Exploring the Small Farm Dream
New England Small Farm Institute
www.smallfarm.org/main/for_new_farmers/exploring_the_small_farm_dream

Soil Texture Calculator
United States Department of Agriculture
www.nrcs.usda.gov/wps/portal/nrcs/detail/soils/survey/tools/?cid=nrcs142p2_054167

USDA Plant Hardiness Zone Map
United States Department of Agriculture
https://planthardiness.ars.usda.gov

Web Soil Survey
USDA Natural Resources Conservation Service
https://websoilsurvey.nrcs.usda.gov/app/WebSoilSurvey.aspx

2: Do It

Online Publications and Videos

Energy Independence: On-Farm Biodiesel Fuel Production
USDA Sustainable Agriculture Research and Education (SARE)
www.sare.org/Learning-Center/Multimedia/Videos-from-the-Field/Energy-Independence-On-Farm-Biodiesel-Fuel-Production

High Tunnel Construction
University of Kentucky
www.youtube.com/watch?v=VPkuRrt1Nw

How to Install Row Covers on Vegetable Crops
University of Minnesota
www.youtube.com/watch?v=X-ko9gT5iA4

Irrigation Water Pumps
North Dakota State University Extension Service
www.ag.ndsu.edu/publications/crops/irrigation-water-pumps/ae1057.pdf

Pumps: How to Select the Right Pump
Irrigation Tutorials
www.irrigationtutorials.com/pumps-selecting-a-pump-step-by-step

Laying Plastic Mulch at Rooster Organics
Abundant Harvest Organics
www.youtube.com/watch?v=G6JW18akeQU

Maintaining Irrigation Pumps, Motors, and Engines
National Sustainable Agriculture Information Service (ATTRA)
https://attra.ncat.org/attra-pub-summaries/?pub=112

Microhydropower Systems
U.S. Department of Energy
https://energy.gov/energysaver/microhydropower-systems

Protecting Water Quality from Agricultural Runoff
U.S. Environmental Protection Agency
www.epa.gov/sites/production/files/2015-09/documents/ag_runoff_fact_sheet.pdf

Thermal Banking Greenhouses
USDA Sustainable Agriculture Research and Education (SARE)
www.sare.org/Learning-Center/Multimedia/Videos-from-the-Field/Thermal-Banking-Greenhouses

Websites

AgriMet
U.S. Bureau of Reclamation
www.usbr.gov/pn/agrimet

Ecological Farming Association
https://eco-farm.org/jobs

Farm Energy Calculators
National Sustainable Information Service (ATTRA)
https://attra.ncat.org/attra-pub-summaries/?pub=304

Farm Energy Efficiency
USDA Natural Resources Conservation Service
www.nrcs.usda.gov/wps/portal/nrcs/main/national/energy

List of Internships
National Sustainable Agriculture Information Service (ATTRA)
https://attra.ncat.org/internships

Sustainable Food Jobs
https://sustainablefoodjobs.wordpress.com

WWOOF-USA
https://wwoofusa.org

3: Sell It

Online Publications and Videos

A Do-It-Yourself Producer's Guide to Conducting Local Market Research
Agricultural Marketing Resource Center
Iowa State University
www.agmrc.org/media/cms/UofGeorgiaorg_7EE4EE6C3DABF.pdf

A Market-Driven Enterprise Screening Guide
Western Extension Marketing Committee
https://pace.oregonstate.edu/courses/sites/default/files/resources/pdf/marketdrivenenterpriseguide2.pdf

How to Start a Successful E-Commerce Business — 6 Tips from Seasoned Pros
U.S. Small Business Administration
www.sba.gov/blogs/how-start-successful-e-commerce-business-6-tips-seasoned-pros

Websites

AMS Organic Prices
US Department of Agriculture
www.ers.usda.gov/data-products/organic-prices.aspx

AMS Wholesale Market Price Reports
US Department of Agriculture
www.ams.usda.gov/market-news/fruit-and-vegetable-terminal-markets-standard-reports

Farm Stay USA
U.S. Farm Stay Association
www.farmstayus.com

Food Hub Center
National Good Food Network
www.ngfn.org/resources/food-hubs

Local Harvest
www.localharvest.org

Outstanding in the Field
www.outstandinginthefield.com

4: Manage It

Online Publications and Videos

Converting Cash to Accrual Net Farm Income
Ag Decision Maker
Iowa State University Extension and Outreach
www.extension.iastate.edu/agdm/wholefarm/html/c3-26.html

Farmer's Tax Guide
United States Internal Revenue Service
www.irs.gov/forms-pubs/about-publication-225

Preparing Agricultural Financial Statements
Northwest Farm Credit Services
http://farmbiztrainer.com/docs/BT_Preparing_Ag_Fin_Statements.pdf

Veggie Compass: A Whole Farm Profit Management Tool
Southern Sustainable Agriculture Working Group
https://www.ssawg.org/gfp-veggie-compass

Your Farm Income Statement
Ag Decision Maker
Iowa State University Extension and Outreach
www.extension.iastate.edu/agdm/wholefarm/html/c3-25.html

Websites

Cost Studies
University of California-Davis, Agricultural and Resource Economics
https://coststudies.ucdavis.edu

Oregon Agricultural Enterprise Budgets
Oregon State University
http://arec.oregonstate.edu/oaeb

Organic Enterprise Budgets
Carolina Farm Stewardship Association
https://www.carolinafarmstewards.org/enterprise-budgets

5: Grow It

Online Publications and Videos

Crop Rotation on Organic Farms: A Planning Manual
USDA Sustainable Agriculture Research and Education (SARE)
www.sare.org/Learning-Center/Books/Crop-Rotation-on-Organic-Farms

Websites

Cover Crop Decision Tools
Midwest Cover Crops Council
http://mccc.msu.edu/selector-tool

Cover Crop Guide for New York Vegetable Growers
http://covercrop.org

Introduced, Invasive, and Noxious Plants
USDA Natural Resource Conservation Service
https://plants.usda.gov/java/noxiousDriver

Organic Fertilizer and Cover Crop Calculators
Oregon State University Extension Service
https://smallfarms.oregonstate.edu/organic-fertilizer-cover-crop-calculator

6: Keep It

Online Publications and Videos

Cottage Food Laws in the United States
Food Law and Policy Clinic, Harvard Law School
www.chlpi.org/wp-content/uploads/2013/12/FLPC_Cottage-Foods-Report_August-2018.pdf

Websites

Affordable Care Act
www.healthcare.gov

AgTransitions
University of Minnesota
https://agtransitions.umn.edu

Business Guide
U.S. Small Business Administration
www.sba.gov/business-guide

Creativity Tools
MindTools
www.mindtools.com/pages/main/newMN_CT.htm

Farm Direct Marketing, Producer-Processed Products
Oregon Department of Agriculture
www.oregon.gov/ODA/shared/Documents/Publications/FoodSafety/FarmDirectMarketingProcessedProducers.pdf

Farmer's Health Cooperative of Wisconsin
www.farmershealthcooperative.com

Farm Link Programs
Farmland Information Center
www.farmlandinfo.org/special-collections/4439

Fair vs. Equal Treatment of Farm Family Heirs
University of Vermont
www.uvm.edu/farmtransfer/?Page=videos.html#fairvsequal

Food Safety Modernization Act
US Food and Drug Administration
www.fda.gov/food/guidanceregulation/fsma/

FSIS Guideline for Determining Whether a Livestock Slaughter or Processing Firm is Exempt from the Inspection Requirements of the Federal Meat Inspection Act
USDA Food Safety and Inspection Service (FSIS)
www.fsis.usda.gov/wps/wcm/connect/16a88254-adc5-48fb-b24c-3ea0b133c939/Compliance-Guideline-LIvestock-Exemptions.pdf?MOD=AJPERES

Niche Meat Processor Assistance Network (NMPAN)
www.nichemeatprocessing.org

State Inspections and Cooperative Agreements
USDA Food Safety and Inspection Service (FSIS).
www.fsis.usda.gov/wps/portal/fsis/topics/inspection/state-inspection-programs/state-inspection-and-cooperative-agreements

State Milk Laws
National Conference of State Legislatures (NCSL)
www.ncsl.org/research/agriculture-and-rural-development/raw-milk-2012.aspx

METRIC CONVERSION CHART

TO CONVERT	TO	MULTIPLY
inches	millimeters	inches by 25.4
inches	centimeters	inches by 2.54
inches	meters	inches by 0.0254
feet	meters	feet by 0.3048
feet	kilometers	feet by 0.0003048
miles	meters	miles by 1,609.344
miles	kilometers	miles by 1.609344
gallons	liters	gallons by 3.785
pounds	grams	pounds by 453.5
pounds	kilograms	pounds by 0.45

INDEX

Page numbers in *italic* indicate photos or illustrations; page numbers in **bold** indicate charts or tables.

A

B

C

D

E

J

K

L

M

N

O

P

Q

R

S

T

U

V

W

Y

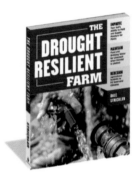

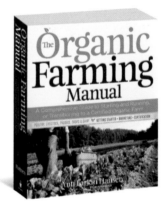